SpringerBriefs in Computer Science

SpringerBriefs present concise summaries of cutting-edge research and practical applications across a wide spectrum of fields. Featuring compact volumes of 50 to 125 pages, the series covers a range of content from professional to academic.

Typical topics might include:

- A timely report of state-of-the art analytical techniques
- A bridge between new research results, as published in journal articles, and a contextual literature review
- A snapshot of a hot or emerging topic
- An in-depth case study or clinical example
- A presentation of core concepts that students must understand in order to make independent contributions

Briefs allow authors to present their ideas and readers to absorb them with minimal time investment. Briefs will be published as part of Springer's eBook collection, with millions of users worldwide. In addition, Briefs will be available for individual print and electronic purchase. Briefs are characterized by fast, global electronic dissemination, standard publishing contracts, easy-to-use manuscript preparation and formatting guidelines, and expedited production schedules. We aim for publication 8–12 weeks after acceptance. Both solicited and unsolicited manuscripts are considered for publication in this series.

**Indexing: This series is indexed in Scopus, Ei-Compendex, and zbMATH **

Francisco S. Marcondes · Adelino Gala ·
Renata Magalhães · Fernando Perez de Britto ·
Dalila Durães · Paulo Novais

Natural Language Analytics with Generative Large-Language Models

A Practical Approach with Ollama
and Open-Source LLMs

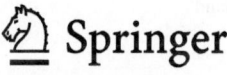 Springer

Francisco S. Marcondes
ALGORITMI Research Centre/LASI
University of Minho
Braga, Portugal

Adelino Gala
ALGORITMI Research Centre/LASI
University of Minho
Braga, Portugal

Renata Magalhães
ALGORITMI Research Centre/LASI
University of Minho
Braga, Portugal

Fernando Perez de Britto
AI Systems Research (AISR)
Investment 4 Impact
São Paulo, Brazil

Dalila Durães
ALGORITMI Research Centre/LASI
University of Minho
Braga, Portugal

Paulo Novais
ALGORITMI Research Centre/LASI
University of Minho
Braga, Portugal

ISSN 2191-5768   ISSN 2191-5776 (electronic)
SpringerBriefs in Computer Science
ISBN 978-3-031-76630-5   ISBN 978-3-031-76631-2 (eBook)
https://doi.org/10.1007/978-3-031-76631-2

This work is financed by National Funds through the Portuguese funding agency, FCT—Fundação para a Ciência e a Tecnologia within project 2022.06822. PTDC. https://doi.org/10.54499/2022.06822.PTDC.

This Springer imprint is published by the registered company Springer Nature Switzerland AG
The registered company address is: Gewerbestrasse 11, 6330 Cham, Switzerland

If disposing of this product, please recycle the paper.

Foreword

This book addresses the subject of qualitative data analysis, a topic of considerable interest within the field of social research. The topic is undoubtedly familiar, given the prevalence of text analysis for the purpose of extracting and analysing emotions. The innovative aspect of this approach is the use of LLMs to extract data from semantic relations, as well as the strategy of using semiotics to support this task. The utilisation of open-source and locally executable LLMs addresses the significant challenge of managing sensitive data, reducing the data breaches and unauthorized access, while also assisting in the mitigation of environmental concerns associated with predictions, as they are designed to operate on business-level computers (often in CPU, eventually on a few GPUs). They also provide the advantage of offline accessibility, flexible customization and the possibility to provide transparency.

From a broader perspective, the book addresses a comprehensive range of topics, encompassing theoretical aspects, business-related issues, legal considerations, and social elements. It should be noted that these topics are not treated in depth, as the intention of this book is to provide a concise and practical overview. In light of the aforementioned, the book has succeeded in its objective of serving as a suitable introduction to both LLM and its utilisation in the extraction of qualitative data. The application of Natural Language Analytics and the proposed outcome of Key Soft Indicators by analysing qualitative data related to climate change, represents a significant advancement in data analysis that may extend beyond the scope of this book.

Overall, the main contribution of this book is to demonstrate how LLMs can be extended beyond their conventional applications, such as translation and summarisation, to more sophisticated tasks that facilitate a deeper understanding of human communication, behavior and decision-making. The byproduct is the utilisation of LLMs integrated with Python to process substantial quantities of data as part of the big data pipeline. It is noteworthy that this is not an entirely optimistic book, as limitations are duly acknowledged and discussed. The discussion prompts readers to consider the current position of LLMs within the hype cycle and to contemplate their future trajectory in terms of development and adoption. It raises crucial questions about the potential cognitive impact of LLMs on the human mind, exploring how these sophisticated models might shape our thinking, learning, and decision-making processes. By examining both the opportunities and challenges presented by LLMs, the discussion encourages a deeper reflection on their long-term implications for human cognition and society as a whole.

I would like to conclude by congratulating the authors for their excellent, concise, and precise work in the highly relevant field of LLMs. Their contribution to the utilization of open-source and locally executable LLMs is commendable. It is impressive how much they have conveyed so effectively in such a limited space.

September 2024

Francisco Herrera
Department of Computer Sciences
and Artificial Intelligence
Andalusian Research Institute in Data
Science and Computational
Intelligence
University of Granada
Granada, Spain

Preface

The inception of this book arose from the authors' collective ambition to explore and utilize generative Large Language Models (LLMs) as "language calculators" to infer and extract soft data, providing semantic analyses and insights expressed in the form of Key Soft Indicators (KSIs). While LLMs are often employed for tasks such as translation and summarization, this book seeks to extend their use to a more sophisticated form called Natural Language Analytics (NLA). This book includes semiotic extraction, sentiment analysis, emotion inference, mindset analysis, and cultural understanding—tasks that enable the processing of complex and multistable textual data to uncover patterns and offer insights into human decision-making.

The goal of this book is to show how generative LLMs can be employed in a downstream process to extract useful data from natural language artifacts in a manner that integrates seamlessly into the data science pipeline, ultimately delivering actionable intelligence to stakeholders. By leveraging these models, analyses can be conducted across a variety of fields, including strategic communication, marketing, human resources, journalism, branding, customer satisfaction, social media, civil defense, risk reduction, and cognitive defense, among others.

This work represents a transdisciplinary endeavor, bringing together expertise from computer science, artificial intelligence, cognitive semiotics, linguistics, management, and information systems, spanning both industry and academia. The convergence of these diverse fields is essential to effectively address the complexities of using LLMs for NLA. Consequently, this book is intended not only for AI specialists but also for professionals across a wide range of fields who are familiar with information systems, decision-making, data science, and natural language processing and who are interested in employing LLMs for natural language analysis, particularly using open-source models on local systems.

Understanding the utility of LLMs as tools for NLA involves recognizing both their strengths and limitations. The book adopts the metaphor of LLMs as "language calculators" to explain how they function by mapping input sentences to output soft data artifacts, driven by complex semantic embeddings. While this metaphor has its limits, it provides a useful framework for understanding the strengths and weaknesses of LLMs in language processing tasks. The book discusses and provides solutions for the inherent variability in LLM outputs and addresses the implications for tasks that require high precision versus those where utility and nuanced interpretation are more critical. Additionally, it tackles challenges related to the ambiguity and variability of natural language and illustrates how LLMs can be effectively set up and applied in this context. A case study is presented on the development of a KSI derived from text data, showcasing real-world applications of these techniques.

Finally, this book does not aim to be an exhaustive guide, but rather to provide practical solutions and guidance based on the authors' experiences. As such, it serves as a starting point for those interested in the evolving field of NLA using LLMs. We hope this book will inspire and equip readers with the tools and knowledge to explore the potential of LLMs in their own work, fostering innovation and advancing understanding in this dynamic area.

Braga, Portugal   Francisco S. Marcondes
Braga, Portugal   Adelino Gala
Braga, Portugal   Renata Magalhães
São Paulo, Brazil   Fernando Perez de Britto
Braga, Portugal   Dalila Durães
Braga, Portugal   Paulo Novais

Acknowledgements

We, the authors, would like to express our deepest gratitude to everyone who has supported us on this journey. A special thanks to our editor Alexandru Ciolan, and his incredible team for their unwavering dedication and expertise. To several academic colleagues, thank you for the invaluable exchange of ideas and insights that have enriched this work. We are also deeply grateful to our families, whose love, patience and encouragement have been a constant source of strength. This book would not have been possible without each and every one of you.

Our hearts and memories will forever hold your support and kindness.

Francisco S. Marcondes
Adelino Gala
Renata Magalhães
Fernando Perez de Britto
Dalila Durães
Paulo Novais

Contents

1   **Introduction** 1
   1.1   What Kind of Tools are the Generative LLMs? 2
   1.2   What Type of Data do Generative LLMs Deliver? 3
   1.3   What are the Applicable Use Cases? 4
   1.4   About this Book 5
   1.5   Book Organization 6
   References 7

2   **Natural Language Analytics** 9
   2.1   NLA Theoretical Inception 12
   2.2   Key Soft Indexes 15
   2.3   Technology Assessment Overview 17
      2.3.1   Model Size 18
      2.3.2   Architecture 18
      2.3.3   Deployment Solution 19
      2.3.4   Optimization Techniques 19
   References 20

3   **Using Ollama** 23
   3.1   Quick-Start 25
   3.2   Using Ollama 26
      3.2.1   Operations on Command Shell 26
      3.2.2   Customized Model Setup 27
      3.2.3   Operations on Python 33
   3.3   Ollama Ecosystem 34
   References 35

4   **Generative Prompt Engineering** 37
   4.1   Generative Prompting Techniques 39
   4.2   Background on Semiotics 41
      4.2.1   Pragmatic Relation Extraction 45
      4.2.2   Theory Underlying Soft Data 45

4.3   Abductive Chain-of-Thought Prompting 45
References 50

5 Case Study: LLM-Based Anxiety Climate Index 53
5.1   Working Dataset 55
5.2   Soft-Data Extraction 56
5.3   Soft-Data Analysis 60
5.4   Soft-Data Indicator 67
References 72

6 Conclusion 75
6.1   Synthesis and Discussion 76
  6.1.1   Parameter Efficient Fine-Tuning 77
  6.1.2   Contextualized Information Enrichment 79
6.2   Related Considerations 79
  6.2.1   Ecological Footprint 80
  6.2.2   Legal Constraints 80
  6.2.3   Economic Effects and Social Impact 81
6.3   Summary and Future Trends 82
References 83

Acronyms

AI	Artificial Intelligence
ACI	Anxiety Climate Index
CSV	Comma Separated Values
DRR	Disaster Risk Reduction
GGUF	GPT-Generated Unified Format
LLM	Large-Language Model
NLA	Natural Language Analytics
NLP	Natural Language Processing
NN	[Artificial] Neural Network
KPI	Key Performance Indicator
KSI	Key Soft Indicator
PEFT	Parameter Efficient Fine-Tuning
UN	United Nations

Chapter 1
Introduction

There is nothing that can be said by mathematical symbols and relations which cannot also be said by words. The converse, however, is false. Much that can be and is said by words cannot successfully be put into equations, because it is nonsense.

Clifford Truesdell, 1966

Abstract This chapter introduces the use of generative large language models (LLMs) in natural language analysis (NLA), focusing on their potential to extract soft data from complex textual information for various analytical purposes. It sets the stage for the book as a practical guide to the application of LLMs in NLA, particularly using open source models on a local system. Given the typical use of LLMs in translation and summarization, it emphasizes the need for a transdisciplinary approach that integrates knowledge from fields such as cognitive semiotics, linguistics, computer science and big data in order to be applied to analytics. Finally, the chapter uses the metaphor of LLMs as "language calculators" to illustrate their function in generating meaningful outputs through semantic embeddings, while also addressing the inherent variability and limitations of these models.

This book was conceived as a result of the authors' efforts to use generative Large Language Models (LLMs) to extract data that can be used for analysis. Instances of them are *relation extraction*, *sentiment analysis*, *emotion inference*, *mindset analysis*, and *cultural understanding* are relevant for processing complex and nuanced textual data, uncovering patterns, and providing insights into human communication, behavior, and social dynamics. This is not yet a widespread task, as generative LLMs are mostly used as assistants for translation, summarization, *etc*. The attempt is to use generative LLMs to extract useful data from natural language artifacts in such a way that it can go through the data science pipeline and deliver intelligence to stakeholders. These analyses can be applied as supportive tools in fields such as *communication, marketing, human resources, journalism, branding, customer satisfaction, reviews, social media, civil defence, risk reduction, cognitive defence*, and others.

F. S. Marcondes et al., *Natural Language Analytics with Generative Large-Language Models*, SpringerBriefs in Computer Science,
https://doi.org/10.1007/978-3-031-76631-2_1

1

This is obviously not straightforward, as it requires a convergence of knowledge from different fields. For this reason, a transdisciplinary effort was undertaken to build the necessary knowledge for this book. Thus, in addition to specialists in computer science and artificial intelligence, the list of authors includes specialists in cognitive science, linguistics, management, and information systems (from both industry and academia). This book is the result of such a convergence of knowledge.

For this reason, it is written not only for artificial intelligence specialists, but also for people in a wide range of fields who are familiar with data science and natural language processing and who want to use LLM to perform natural language analysis (NLA) locally with open source LLMs. Above all, this is a practical book[1] , describing how the authors have used LLM to meet specific needs and goals, which may be useful to other people with similar needs.

For a reference, perhaps the most common application of NLA is in recommendation systems, where information such as entities and sentiments are often extracted to support all sorts of decision-making [1–4]. For example, it might be relevant to understand whether a negative text (e.g. a complaint) comes from a friendly (agreeableness) or a nervous (neuroticism) stance [5]. As well as providing a fine-grained understanding, it also helps to provide the most appropriate response to the customer's needs (including cognitive ones [6]). Nevertheless, possibilities for using NLA include predicting national elections [7], student grades without academic data [8], the sentiment that may arise on people answering a Likert questionnaire [9], *etc.*

1.1   What Kind of Tools are the Generative LLMs?

Perhaps a good metaphor for describing LLMs is as *language calculator,* i.e. a calculator that computes language sentences instead of mathematical expressions. Like any metaphor, this is a short-lived one, but it is suggestive enough to illustrate its strengths and weaknesses. Nonetheless, when it comes to prediction, it is a reduction procedure (i.e. a calculation) that maps an input sentence into an output sentence.

To elaborate a little [10], an LLM recursively predicts the "next" token based on the context composed of the input sentence (called the prompt) and the tokens already generated; it stops at the prediction of a termination token. Note also that the tokens are not actually words, but a vectorial representation of them (called embeddings). In short, the embedding space of a word is computed based on its colocation with other words, considering numerous texts. The dimension number of a word embedding varies according to the model but, currently, often ranging between 768 and 4096 [11]. This is used for capturing the *semantics* of the word.

Semantics, morphology, syntax, and pragmatics, are the constituents of a *language* [12]. A programming language compiler can also be considered a language calculator, but focused on the syntactical level. Beyond syntax, semantics explores the meaning of words and sentences, while pragmatics focuses on how sentences are

---

[1] For an in depth theoretical exposition, refer to https://web.stanford.edu/~jurafsky/slp3/.

used in specific communicative contexts. In addition, *discourse* involves the study of whole texts, analyzing the relationships between sentences and text structures. Then, the calculation aims to predict token co-locations based on semantic features.

It can be argued that the language calculator metaphor fails insofar that the same input can lead to different outputs. That variation, however, is the result of a pseudo-random algorithm used to select the best matching tokens. Therefore, by setting the seed, the output become deterministic (to be further discussed in Chap. 3). The main difference between an LLM and a calculator is that the behavior of the latter is based on predefined rules, whereas the behavior of the first is based on unsupervised learning. Nevertheless, after training, the behavior of an LLM is also based on a set of rules (devised during training). In this sense, an LLM will "never" select as the next word a token with a low co-location probability but the average use of such a word. Recursively, the average use of sentences, paragraphs, *etc.* In other words, it will reproduce, in a more or less conservative way, the "common sense" of the data set (including biases and prejudices).

1.2   What Type of Data do Generative LLMs Deliver?

In summary, LLMs are driven by embedded representations of the semantics of language, where words are encoded as collections of "semantic tokens" learned from a large textual dataset. These tokens capture complex semantic relationships in which linguistic elements co-occur. An example is the word *bank*, which occurs in the context of both a river and money. This results in an "embedding spectrum" which is tuned (or nuanced) by the LLM according to the context given in the prompt. When translating or paraphrasing, the models can then map "semantic tokens" across languages or phrases while preserving the underlying "numerical meaning".

The architecture behind the LLMs is called Transformers [13], which is divided into *encoder* and *decoder* parts. Generative LLMs, such as GPT, are often decoder-only, whereas classification LLMs, such as BERT, are often encoder-only. NLA is then usually done with encoders and not with decoders as suggested in this book. The option of choosing decoders came from the observation that the performance of BERT-like models is often not significantly better than other approaches, for the sentiment analysis task comparing the results of VADER and BERT see [14–18]. On the other hand, by asking a GPT to classify the sentiment of words and comparing it with VADER's human annotation, the predictions fall within the standard deviation [19]. They are then suitable for the sentiment classification purpose.

Another important difference between BERT-like and GPT-like models is that the former only returns embeddings, whereas the latter returns words. The consequence is that the former needs to go through a training process to tune it to the intended downstream task, whereas the latter does not. This implies the need for a dataset to train the former (which may not be available), whereas the latter will provide an answer even for unknown tasks (albeit incorrectly). Since the performance of the former is not that superior, for some given situations it may be best suited to produce a prompt and a setup that induces the latter to provide sufficiently accurate responses than to build a dataset from scratch for training the first.

Such an approach is to be considered in terms of *utility* [20] rather than precision. Utility is a concept used in economics, and particularly in game theory, to refer to the value that an agent derives from an event. An event *increases utility* if it provides value to the agent. Note that accuracy only increases utility up to a certain point. For example, even though it is possible to calculate a percentage with many decimals, they are often rounded to one or two digits (this would be the point of maximum utility). Words, phrases, and texts act as semantic packages, each carrying a range of potential interpretations. Then, ultimately it is not possible, even for a human, to know for sure the sentiment of a phrase (that's why annotation is carried out by several people in the search for an average [21]), even worse for a machine [22]. Nevertheless, sentiment analysis produces utility for stakeholders.

In this sense, there are *hard data*, i.e. a measurement of an objective phenomenon, e.g. the number of products manufactured in a day; and *soft data*, i.e. a measurement of a subjective phenomenon, e.g. the level of satisfaction of a customer. Unlike hard data, which provides precise and often numerical values, soft data encompasses the qualitative dimensions, in the case of this book, of language and communication. Because it is subjective, variability, in the sense that measuring the same event can yield different values, is inherent in soft data. It is therefore not unexpected that a decoder will produce different labels when asked to classify the same input into different random states. Despite the variability, it is not expected that the labels are inconsistent or incompatible (the prompt must be properly crafted for a task given the tendency of the model, refer to Chap. 4), but that they explore different nuances of the input text.

1.3   What are the Applicable Use Cases?

From an enterprise perspective, the adoption of nuanced metrics (soft data) lies beyond traditional Key Performance Indicators (KPIs). Decision-making is heavily influenced by quantitative metrics (hard data). They provide a quantitative measure of performance across various organizational facets. From sales figures to operational efficiency, KPIs have been instrumental in setting targets and measuring progress. They are straightforward, data-driven, and primarily focus on the outcome. However, as businesses evolve, the realization that numbers alone cannot capture the full spectrum of organizational performance has come to the fore. This is where Key Soft Indicators (KSIs) step in. KSIs reflect an understanding that language and human interactions are critical components of performance measurement, offering interpretations that traditional metrics may overlook, spotlighting the soft aspects that drive performance. Unlike KPIs, KSIs focus on the qualitative aspects of performance, where linguistics nuances can be measured.

NLA systems attempt to extract data from natural language to support decision making. However, the inherent ambiguity and variability of natural language presents challenges in extracting accurate and consistent information that can be used in a quantifiable way. LLMs are then used to extract meaningful insights from unstructured text data, recognizing the variability of the result as a property. For example, intent analysis can be used to identify the underlying goals or objectives expressed in customer support requests or product reviews. Topic analysis can automatically categorise documents or conversations into relevant themes or topics, facilitating content organization and retrieval. Aspect analysis can analyze product reviews or feedback to identify specific attributes or features discussed, enabling targeted improvements.

Sentiment and emotion analysis can measure the overall sentiment or emotional tone expressed in textual data, helping organizations monitor customer satisfaction or public opinion. Mindset analysis can reveal the underlying beliefs, attitudes or cognitive frameworks reflected in the language used by an individual or group. Social analysis can examine language patterns to infer social dynamics, power relations or cultural norms within a given context. Rhetorical analysis can assess the persuasive strategies, argumentative techniques or stylistic devices used in written or spoken communication.

By combining these different natural language processing techniques, NLA systems can perform more sophisticated analyses that take into account the nuances, complexities and multifaceted nature of human communication. By integrating techniques such as sentiment analysis, emotion analysis and mindset analysis, it is possible to create a detection system that can help monitor mental health risks and provide timely intervention. Furthermore, by combining social analysis and sentiment tracking over time, policymakers and public health organizations can gain insights into shifting societal attitudes, emerging concerns, or demographic disparities related to mental health. This can inform the development of more inclusive and effective mental health policies, resource allocation strategies, and community outreach programs.

1.4 About this Book

Above all, LLMs are a fairly new technology, so we are still learning how to use them properly. In addition, new ideas and artifacts are being delivered on a daily basis, and it is difficult to keep track of all this material and select from the many simultaneous proposals. Thus, the expectation in this book is not to exhaustively cover all the possibilities, nor the latest technology, but, as mentioned, to report how the authors experience the usage of LLMs to perform Natural Language Analysis (NLA) at an industrial level. As this is an early area of knowledge, it is not possible to claim that this is the most effective use, but that it is useful for the authors and, hopefully, for other people with similar needs.

It is not the intention of this book to present a complete product in any sense, but to explore and examine how to use Large-Language Models (LLMs) to obtain data from natural language artifacts. It highlights tools, techniques, and principles that can be used for analysis in any project. For that purpose, a case study is proposed as an example, with the intention being to provide an underlying object that links the various elements of discussions.

This book also does not cover the full range of NLA applications, but provides insights into some of them throughout the chapters. Furthermore, although this is a book about data analysis, it assumes that the reader is familiar with common tools such as the pandas and seaborn modules, as well as data wrangling techniques commonly used in Natural Language Processing (NLP) such as NLTK and SpaCy.

Finally, the idea of this book is to use LLMs in the context of personal computers, meaning that everything is expected to run locally, without the need for expensive GPUs or any third-party infrastructure. Note that there is no magic: depending on the processing volume and computer power, the processing would take hours or days to complete. However, the idea is that the topics in this book can hopefully be done in a feasible amount of time. Apart from the economic and time aspects, this solution is also important because it ensures that sensitive data is not shared with unknown parties. The trade-off is that fine-tuning is beyond the scope of this book, but a curious reader can refer to the PEFT repository on GitHub (https://github.com/huggingface/peft).

1.5 Book Organization

This book sequentially explores the expansive domain of generative LLMs, starting with foundational concepts and advancing through practical applications and advanced methodologies. From the initial overview of LLMs and their data generation capabilities, to the in-depth examination of Natural Language Analytics (NLA) and its real-world impact, each chapter has built upon the last to create a cohesive understanding.

Generative prompt engineering and the use of tools like Ollama have demonstrated how theoretical knowledge can be applied in practice, while case studies have illustrated the complexities and solutions in real-world applications. The readers are expected to learn from the advanced topics how to push the boundaries of what is possible with generative LLMs. By synthesizing these insights, the final chapter offers a comprehensive view and highlights future trends, paving the way for ongoing exploration and innovation in this dynamic field.

For a specific description of each chapter,

Chapter 1 sets the stage for the book by presenting an overview of generative Language Models (LLMs), discussing their tools, the type of data they deliver, and their diverse applications. It provides foundational knowledge necessary for understanding the depth and breadth of generative LLMs, essential for exploring their practical applications and theoretical implications.
Chapter 2 explores the realm of Natural Language Analytics (NLA), emphasizing its theoretical inception, technology assessment, and key soft indexes. This chapter highlights the role of NLA in analytics and decision-making processes,

demonstrating state-of-the-art applications and the potential insights that can be derived from NLA.

Chapter 3 serves as a practical guide to using Ollama, a tool designed for working with generative LLMs. This chapter covers the entire range from quick-start procedures to model setup and integration with Python. This chapter provides the necessary technology background for properly understand and running open-source LLMs locally.

Chapter 4 is dedicated to the techniques and principles of generative prompt engineering, this chapter introduces various prompting techniques and the semiotic background essential for effective prompt creation. It delves into semiotic prompting strategies to enhance the efficacy and relevance of system interactions using generative LLMs.

Chapter 5 uses climate change as a case study to explore the practical application of Natural Language Analytics by extracting insights from textual data. Through practical examples, including the creation and processing of a dataset and customization of Ollama models, the chapter illustrates the methods used to generate an Anxiety Climate Index (ACI).

Chapter 6 synthesizes the information presented throughout the book, providing a cohesive summary of the key concepts, methodologies, and applications. It discusses the overall impact of these technologies (including some of their opportunities and threats) then highlights future trends and potential areas for further research and development.

References

1. S. Algarni and F. Sheldon. Systematic review of recommendation systems for course selection. *Machine Learning and Knowledge Extraction*, 5(2):560–596, 2023.
2. G. Bathla, P. Singh, S. Kumar, M. Verma, D. Garg, and K. Kotecha. Recop: fine-grained opinions and sentiments-based recommender system for industry 5.0. *Soft Computing*, pages 1–10, 2023.
3. D. Roy and M. Dutta. A systematic review and research perspective on recommender systems. *Journal of Big Data*, 9(1):59, 2022.
4. S. Wang, Y. Wang, F. Sivrikaya, S. Albayrak, and V. W. Anelli. Data science for next-generation recommender systems. *International Journal of Data Science and Analytics*, 16(2):135–145, 2023.
5. R. R. McCrae and O. P. John. An introduction to the five-factor model and its applications. *J. Pers.*, 60(2):175–215, 1992.
6. V. Joines and I. Stewart. *Personality Adaptations: A New Guide to Human Understanding in Psychotherapy and Counselling*. Lifespace, 2002.
7. R. Martins, J. J. Almeida, P. Henriques, and P. Novais. Predicting an election's outcome using sentiment analysis. In Álvaro Rocha, Hojjat Adeli, Luís Paulo Reis, Sandra Costanzo, Irena Orovic, and Fernando Moreira, editors, *Trends and Innovations in Information Systems and Technologies*, pages 134–143, Cham, 2020. Springer International Publishing.
8. B. Lacerda, H. Marcondes, F. S.and Lima, D. Durães, and P. Novais. Prediction of students' grades based on non-academic data. In Marcelo Milrad, Nuno Otero, María Cruz Sánchez-Gómez, Juan José Mena, Dalila Durães, Filippo Sciarrone, Claudio Alvarez-Gómez,

Manuel Rodrigues, Pierpaolo Vittorini, Rosella Gennari, Tania Di Mascio, Marco Temperini, and Fernando De la Prieta, editors, *Methodologies and Intelligent Systems for Technology Enhanced Learning, 13th International Conference*, pages 87–95, Cham, 2023. Springer Nature Switzerland.

9. R. Magalhães, F. S. Marcondes, D. Durães, and P. Novais. Emotion extraction from likert-scale questionnaires. In Paulo Quaresma, David Camacho, Hujun Yin, Teresa Gonçalves, Vicente Julian, and Antonio J. Tallón-Ballesteros, editors, *Intelligent Data Engineering and Automated Learning – IDEAL 2023*, pages 166–176, Cham, 2023. Springer Nature Switzerland.

10. D. Jurafsky and J. H. Martin. *Speech and Language Processing*. draft (url-https://web.stanford.edu/ jurafsky/slp3/), third edition, 2023.

11. N. Muennighoff, N. Tazi, L. Magne, and N. Reimers. Mteb: Massive text embedding benchmark. arXiv preprint arXiv:2210.07316, 2022.

12. C. Rühlemann. *Corpus linguistics for pragmatics: A guide for research*. Routledge, October 2018.

13. A. Vaswani, N. Shazeer, N. Parmar, J. Uszkoreit, L. Jones, Aidan N. Gomez, Ł. Kaiser, and I. Polosukhin. Attention is all you need. *Advances in neural information processing systems*, 30, 2017.

14. T. Fontes, F. Murçós, E. Carneiro, J. Ribeiro, and R. J. F. Rossetti. Leveraging social media as a source of mobility intelligence: An nlp-based approach. *IEEE Open Journal of Intelligent Transportation Systems*, 2023.

15. J. Majidpour and K. Al-Barznji. Opinions for receiving covid-19 vaccines based on sentiment analysis. *Journal of Pharmaceutical Negative Results*, pages 1648–1659, 2022.

16. A. J. Nair, G. Veena, and A. Vinayak. Comparative study of twitter sentiment on covid-19 tweets. In *2021 5th International Conference on Computing Methodologies and Communication (ICCMC)*, pages 1773–1778. IEEE, 2021.

17. D. Waterman. *Predicting Twitter Sentiment on the Russo-Ukrainian was using Lexicon-based and Transformer Models*. PhD thesis, Tilburg U., 2022.

18. T. Xie, Y. Ge, Q. Xu, and S. Chen. Public awareness and sentiment analysis of covid-related discussions using bert-based infoveillance. *AI*, 4(1):333–347, 2023.

19. F. Marcondes, A. Gala, M. Rodrigues, J. J. Almeida, and P. Novais. Lexicon annotation with llm: a proof of concept with chatgpt. Paper submitted to a Conference., 2024.

20. S. Mukherjee, M. Mukherjee, and A. Ghose. *Microeconomics*. Prentice Hall India Pvt., Limited, 2004.

21. S. Mohammad. Best practices in the creation and use of emotion lexicons. In Andreas Vlachos and Isabelle Augenstein, editors, *Findings of the Association for Computational Linguistics: EACL 2023*, pages 1825–1836, Dubrovnik, Croatia, 2023. Association for Computational Linguistics.

22. A. Tarski. The semantic conception of truth: and the foundations of semantics. *Philosophy and Phenomenological Research*, 4(3):341, 3 1944.

Chapter 2
Natural Language Analytics

I like words more than numbers, and I always did.

Paul Halmos, 1985

Abstract This chapter discusses the foundational theory and practical applications of language models in the realm of natural language analytics (NLA). It explains that understanding semantic relationships, clustering similar words, or inferring sentiment based on word proximity embed the subjective and qualitative nature of language, categorized as soft data, which may be pivotal for nuanced decision-making in organizations. Key Soft Indicators (KSIs) are introduced as a tool that quantify soft data like customer satisfaction or organizational culture, complementing the more traditional, quantifiable Key Performance Indicators (KPIs). Furthermore, the chapter discusses the integration of advanced computing techniques to reduce data and analysis latency, enhancing the speed and efficiency of decision-making processes in big data environments.

The application of analytics in business is widespread in various fields [1, 2] including but not limited to manufacturing, logistics, financial markets, supply chains and sales [3–6]. In general, it is a mode of unlocking intelligence [7] based on data. The employment of big data as part of an enterprise information system depends on an organizational culture focused on defining actions and making decisions through constant data and information analysis [8]. By way of historical background, the application of analytics in business began in the mid-1950s with the development of Statistical Analysis Systems (SAS). In the 2000s, the volume of data increased significantly, leading to technologies such as Hadoop and NoSQL databases to support data warehouses and balanced scorecards. It was not until the 2010s that AI became widely used in the field, paving the way for the notion of *big data* [9]. These are the three ages of analytics *cf.* [2].

Netflix, for example, is a business built on the use of data analytics. Its primary application is to understand its customers' preferences by analysing their behavior in relation to selected films. This analysis not only provides personalized recommendations tailored to each customer's profile, but also informs decisions about the purchase of licenses based on predicted screen views (i.e. helping to predict a price equilibrium range and the return on overinvestment). Ultimately, all the BigTechs (e.g. Netflix, Meta, Google, *etc.*) are highly dependent of analytics and its power, then also a good example for the power of, analytics.

In a broader context, cloud computing has made data storage and processing more scalable and accessible, while real-time analytics has become increasingly important for businesses to respond quickly to market changes. In addition, the integration of Internet of Things (IoT) devices has provided even more data sources for analysis, expanding the possibilities of what can be measured and optimized. Privacy concerns and regulations, such as General Data Protection Regulation (GDPR), have also become more prominent, impacting the way businesses handle and process data.

In the context described, analytics can be defined as:

Analytics
Systematic computational analysis of data to guide decisions and actions.

For a reference, other equally suitable definitions are:

- The extensive use of data, quantitative and statistical analysis, explanatory and predictive models, and fact-based management to guide decisions and actions [8]; and
- A process of transforming data into actions through analysis and insights in decision-making and problem-solving [10].

These definitions all suggest that supporting decision-making is the goal of analytics [11]. A decision is a process aimed at choosing courses of action in the light of existing interests and goals [12]. Decisions are related to different issues and have different levels of uncertainty, bias, and impact. Uncertainty is often due to factors such as abstraction, complexity, scope, knowledge of alternatives and associated probabilities [9]. Note that the decision process involves criteria and sub-criteria, tangible or intangible, used to rank alternatives according to a hierarchy established by the decision maker [13]. Finally, decisions are certainly not driven by data alone, but data is at the core of any *rational decision*.

A cognition that is also in the core of the decision process is the *insight* [14]. Insight is a data-driven cognition that involves awareness of the relationships between considered objects and their properties, leading to a resolution [15]. Despite varying conceptualizations of insight across fields such as psychology, psychiatry, communication, marketing, and management, the Gestalt-based definition aligns well with the application of analytics in organizational decision-making processes.

The value of an insight, despite the data on which it is based, cannot be assessed directly [16], but it can be assessed by its outcome in terms of improved organizational intelligence and decision making [17] or cost reduction and revenue increase [18]. Data can be either "hard" (objective or measurable, e.g. a company's revenue) or "soft" (subjective or intangible, e.g. brand perception); both are equally important in guiding the decision process. Because of their subjectivity, soft data are inherently multistable (the rater is subject to spontaneous subjective changes in the observed phenomenon [15]) in the sense that the same events may be evaluated differently. Then, either the human rater or the LLM is likely to provide one possible interpretation. One field of knowledge that has emphasis on the dynamics of language and thoughts is *psycholinguistics* [19, 20]. It is an intersection of psychology, physiology, and linguistics that explores the interplay between neural structures and linguistic functions (acquisition, comprehension, production, and development), ranging from experimental psychology to neuroimaging [21]. In this sense, in addition to its relation to soft data, the subjective value embedded in linguistics is intrinsic to human decision-making processes, thus an object to be mapped and explored.

Following these notions, soft data can be defined as:

Soft Data
Qualitative and multistable aspects of a phenomenon, often expressed as a category.

Perhaps the richest source of soft data is text data, such as customer feedback, complaints, social media interactions, *etc.*. As these are language artifacts, the idea of using LLMs to extract soft data from them is straightforward. Then, for example, depicted in Fig. 2.1, an LLM is asked to identify the key topic and the associated sentiment, emotion and evoked feeling from the first 5 film descriptions in the IMDb dataset. Since these models are programmed to deliver slight different replies, for a thorough understanding the same prompt was run twice. Note that the data presented is partially pre-processed in csv format (for reference, the 2.1 and 2.2 prompts illustrate raw outputs of plain prompts), but the text of the cells is raw, so it is possible to clean the data using traditional NLP and treat it as any other categorical dataset.

This then leads into another definition:

Natural Language Analytics
Insights gained from *soft data* extracted from text as an element of *analytics*.

In other words, Natural Language Analytics (NLA) aims at mapping and analyzing the relational dynamics between words, phrases, and contexts, and it provides a nuanced and quantifiable understanding of both explicit and latent information within language. This level of analysis is instrumental in setting the path for Key Soft Indicators (KSIs). KSIs represent a class of metrics that move beyond traditional numerical data to encapsulate the subtleties of human expression and communication. These indexes are crucial in areas where the sentiment, emotional tone, mindset, cultural context, or implied meanings are significant for understanding and quantifying the underlying messages conveyed through language expressed in the form of communication flow.

```
1-Redemption Through Decency; Positive; Hopefulness; Empathy
2-Organized Crime Dynasty Succession; Negative; Tension/Anticipation; Loyalty vs. Ambition
3-Gotham's struggle against the Joker; Negative; Anxiety, Fear; Determination, Resilience
4-The Rise of a Crime Syndicate; positive; Empowerment, ambition; Determination, visionary leade
5-Jury Deliberation Resistance; Negative; Frustration; Justice Sense
```

(a) run #1.

```
1+Redemption; Positive; Empowerment; Compassion
2+Succession in Organized Crime; Negative; Resistance, Distrust; Loyalty, Determination
3+The Joker's threat to Gotham; Negative; Anxiety/Fear; Determination, Resilience
4+Vito Corleone's early life; positive; admiration/inspiration; resilience/strength
5+Jury deliberation resistance; Negative; Anger/Frustration; Perseverance
```

(b) run #2.

Fig. 2.1 Comparison between the data extracted with two runs of phi3:instruct from the overview of 5 movies in the IMDb dataset

In a real-world scenario it is also worth of highlighting the relationship between time and business value regarding the decision-making [22]. A decision process is triggered by a business event requiring a response and ends with an action taken in response. There are two intermediate steps of interest: making data available for analysis and making analyses available for decision-making. The time required for performing the first is called *data latency* (time to collect, prepare, and store data) and the second is called *analysis latency* (time to access, analyze, and generate information and alerts). The issue is how swiftly an organization can turn raw data into actionable insights.

2.1 NLA Theoretical Inception

Put simply, a language model is the result of applying a machine learning algorithm to a text to find patterns of relationships between words (e.g. co-locations, co-occurrences, etc.) [23]. The Hidden Markov Model (HMM) is perhaps the simplest example, working by establishing probability relationships between words. A more sophisticated model is based on using an artificial neural network (NN), which is trained by predicting the next word and back-propagating the errors. The resulting weights associated with each word is that word's embedding. A *word embedding* is then a vector of a multidimensional plane. Since the word embedding is based on the prediction of a word given a context, it captures both syntactic and semantic patterns.

Using a dimensionality reduction algorithm, it is possible to reduce a word embedding to a two-dimension vector that can be plotted to gain insight. Three plot examples of embeddings as such are depicted in Fig. 2.2. For a brief explanation, embeddings are mostly based on co-locations, which in turn are based on the idea that similar words occur in matching contexts [23]. For example, as suggested by Fig. 2.2a and as would be expected, the distance between man and king and woman and queen is similar. In another example, as suggested by the Fig. 2.2b and as would also be expected, it is possible to cluster semantically related words.

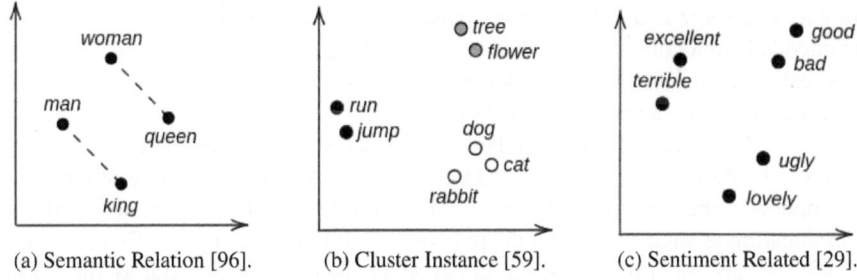

(a) Semantic Relation [96].   (b) Cluster Instance [59].   (c) Sentiment Related [29].

Fig. 2.2 Examples of word-embeddings

Following the other two, Fig. 2.2c suggests that it is possible to infer the sentiment of a word based on operations on the distance from other words [24]. That is, the sentiment of a word is given by the context in which a word appears in the training set. For example, consider the word dog, although it may appear in contexts with both ugly and lovely (Fig. 2.2c), in the general case it can be expected that the co-occurrence between dog and lovely will be higher than between dog and ugly. In this case, the distance between dog and lovely would be smaller than that between dog and ugly. Suppose a sentiment classifier that predicts lovely as positive would imply that dog is positive. This is obviously a simplification, but it touches on the main idea.

The limitation of this approach is that embeddings as such are fixed. Consider again the word dog; in the context of someone referring to their pet, it would probably be positive, but in referring to someone as a dog it is more likely to be negative. A feature of the Transformer architecture[1] is, given the context window, to adjust the embedding of a word through an encoding process (i.e. positional encoding, attention mechanism, *etc.*), and then cope with situations as such. In this case, the autoregressive operation of the model would help it to identify the sense in which the word dog is being used.

The Transformer architecture is composed by two parts: *encoder* and *decoder*. These parts can be used together or spared. **Generative LLM refers to a decoder-only Transformer**. Roughly, this is done by manipulating three scores: query, key and value. The query is the score of the word being processed, which is compared with the key of all the preceding words. This results in the "distance" between the current word and all previous words, normalized between 0..1. These numbers are

---

[1] It is not the intention of this book to delve into theoretical issues. However, the curious reader may find the links of interest:

- For an informal but rigorous presentation:
  https://www.youtube.com/playlist?list=PLblh5JKOoLUIxGDQs4LFFD--41Vzf-ME1
- For a formal and comprehensive presentation:
  https://web.stanford.edu/~jurafsky/slp3/.

used to scale the value of all words (including the one being processed), resulting in an encoded embedding vector. This vector usually passes through a fully connected NN (called *head*) which will predict the next word in the sentence.

Then, if the "distance" between dog and pet is smaller than between dog and person, the model is likely to embed dog closer to positive words. Thus, the dog is likely to be classified as positive (see Prompt 2.1). Similar reasoning can be applied to the annotation of any word, phrase or text. It is worth noting that a classification as such will reflect the patterns found in its training set. Therefore, the result will vary according to the culture in which the texts were produced, the biases of the training data, *etc.* Issues as such is further discussed in Chap. 4.

Prompt 2.1 phi3:instruct (run #1)

```
1  >>> classify "dog" into positive or negative
2  The term "dog" is generally perceived as positive. Dogs are often
       ↪ associated with loyalty, companionship, love, and joy in
       ↪ many cultures around the world. They play an important
       ↪ role in various human activities such as therapy, service
       ↪ (e.g., guide dogs), and working roles (e.g., herding
       ↪ livestock or search and rescue operations). However, it's
       ↪ worth noting that individual experiences may vary, but
       ↪ overall, "dog" is viewed positively by the majority of
       ↪ people.
```

It is worth mentioning that chat-based LLMs are not stable in the sense that they can result on different predictions for a same prompt (this issue is addressed in Chap. 3). In short, this behavior is due to a random selection of best matching tokens delivered by the head of the model, for a reference compare prompts 2.1 and 2.2 where a same prompt results on two different classifications.

Prompt 2.2 phi3:instruct (run #2)

```
1  >>> classify "dog" into positive or negative
2  The word "dog" is neutral and not inherently positive or negative
       ↪ . However, whether it's perceived as positive or negative
       ↪ can depend on the context and personal experiences of an
       ↪ individual. For example:
3
4  Positive contexts:
5  - A pet dog bringing joy to its owner
6  - Dogs being used in therapy for emotional support
7  - The loyalty and companionship that dogs offer
8
9  Negative contexts:
10 - Fear or phobia related to dogs (though this is more about the
       ↪ individual's perception)
11 - Negative experiences with aggressive dogs
12
13 In general, "dog" by itself does not classify as positive or
       ↪ negative.
```

Note that both responses (positive and neutral) cannot be said to be wrong but valid perspectives for that analysis. In other words, it is the qualitative evaluation of a subjective phenomenon, therefore an instance of *soft data* (refer to Sec. spsexrefsec:data). As a matter of result, despite the responses are different, they point to the same direction: "it depends on the individual experience with dogs". From a decision-making perspective, both responses provide data that would yield to useful *insights*, fulfilling its business purpose.

The advantage (but also the weakness) of using a decoder-only transformer for NLA is that it uses the model that has already been trained and does not need to be fine-tuned in the downstream task before being used. This property must be taken into account when considering the data produced by these models. As this is particularly valuable in the absence of a data set, a validation set may not be available for evaluation (in some cases it may not even make sense). Performance evaluation from a sample relies on a specialist (e.g. a psychologist, a cognitive scientist, etc.) who is able to take into account the natural variability of soft data. There are ways to improve the results, as discussed in the following chapters, but human evaluation for validating the results is unavoidable.

2.2 Key Soft Indexes

Key Performance Indicators (KPIs) play a crucial role in analytics, since they are often based on hard data. Similarly, Key Soft Indicators (KSIs) are used to refer to indicators based on soft data. Such a separation is necessary as they serve to different purposes.

In short, KPIs are traditional metrics used extensively in business and operational environments to measure clear, quantifiable outcomes such as financial performance, operational efficiency, or market penetration. These are typically hard data points derived directly from numerical analytics, which help organizations gauge their performance against specific objectives. On the other hand, KSIs are derived from soft data and focus on quantifying the nuances in qualitative data.

Unlike the direct numerical output seen with KPIs, KSIs aim to provide measurable insights into less tangible aspects such as customer satisfaction, organizational culture, mental health, or public sentiment. These soft data indicators help quantify abstract concepts by applying analytical rigor to text and language data, translating subjective content into quantifiable metrics. KSIs therefore extend the analytical capability to encompass aspects of business and communication that KPIs may not capture, such as the tone of customer feedback, trends in employee communication, or shifts in public opinion. This quantification of qualitative data is a renewed form of analysis, bridging the gap between numeric and language analysis.

NLP tasks, currently mostly supported by LLMs, are cornerstone to such effort. For example, Named Entity Extraction (NER) is necessary for organizing and retrieving data by identifying and categorizing elements like geographical locations or organizational names. Topic Analysis allows researchers to detect prevalent themes within large text corpora, facilitating content classification and research trend identification. Intent Analysis and Aspect Analysis further dissect texts to ascertain the communicative intentions and to evaluate specific aspects such as product features in reviews. Consider, for an instance, Fig. 2.1. Without the support of NLP it is not possible to assess the values in order to derive a KSI.

As an illustration, still looking at the same figure, it can be said that the emotional valence of the films does not seem to influence the perceived enjoyment of it. On the other hand, "determination" and related feelings such as "resilience" are common in these films. Note also that the emotions of "frustration", "anxiety" and "fear" are related emotions that are expected to appear in the context of "determination" and "resilience". Obviously, no conclusions can be drawn from these data, but some insights into the stories of successful films begins to emerge.

Soft data includes then emotion, sentiment, and various forms of attitude analyses that can be used for analytics. In the case of NLA, emotion analysis, for instance, is used in psychological studies to assess the expression of feelings in text, enhancing understanding of human affective states. Sentiment Analysis categorizes the emotional tone of texts, an important tool in sociolinguistic research. Furthermore, analyses like Mindset and Social Attitude reveal underlying cognitive frameworks and societal norms, respectively, providing insights that are valuable in cultural studies and social sciences.

Quantitative tasks within NLA involve methods such as *frequency analysis*, which is used to identify prevalent terms or concepts within a dataset. *Distribution and variance analyses* are used to understand the spread and variability of data points, applicable in statistical linguistics and probability studies. *Correlation analysis* helps determine relationships between linguistic features, often used in computational linguistics to explore syntax or semantics relationships. These quantitative methods are foundational in transforming raw text data into structured, analyzable datasets capable of providing valuable insights and support decision-making.

One strength of NLA is to provide quantitative data extracted from nuanced linguistic analysis. For example, merging sentiment and frequency analyses, provides a layered understanding of how public sentiment towards certain topics evolves over time, capturing shifts in societal attitudes that might correlate with major events or policy changes. Additionally, combining Named Entity Recognition with Categorical Data Analysis can suggest relationships between various entities such as organizations, locations, and personal names, and categorical outcomes like sentiment scores or thematic groupings, offering invaluable insights for research in fields like political science, cultural studies, or urban studies.

Common analytics (with a possible set-up) includes:

Detection   For applications like detecting phishing attempts, a 70B model can identify subtle nuances in language more effectively than smaller models, although it may incur a cost in terms of slower response times due to its larger size. The trade-off here is between accuracy in detection and the speed of response, which is critical in cybersecurity environments.

Alert   In scenarios where response speed is paramount, such as real-time financial or security monitoring systems, an 8B model might be more advantageous. Its smaller size facilitates faster processing, which is crucial for triggering timely alerts in dynamic and high-stakes environments.

Insight   A 27B model strikes a balance between computational demands and depth of analysis, making it well-suited for generating detailed market analysis or consumer insights where both thoroughness and speed are valued.

As a summary, binding advanced NLP tasks with foundational analytical techniques, researchers can leverage this synergy to conduct intricate analyses, for instance identity and quantify phishing attempts, identify quantify and alert for risks, or identify and quantify topics and aspects of the market to obtain insights. The outcome expand decision-making possibilities through insights and facilitating more informed conclusions. This approach underscores the evolving landscape of NLA and the path for building Key Soft Indicators.

2.3   Technology Assessment Overview

Latency form a comprehensive framework for evaluating the timeliness and effectiveness of decision-making processes in organizations, especially in the context of big data and advanced analytics [25]. The lower the latency, the better the efficiency. This in turn depends on the computing power, data architecture and analysis tools used. Therefore, when using LLMs, it is necessary to consider their performance in terms of cost/benefit, i.e. the search for a balance between adequate performance, the cost of the necessary infrastructure and the ability to deliver timely results. Once the ROI and future returns have been estimated on the basis of these factors, they should be weighed against the associated risks for a comprehensive evaluation [25]. Therefore, when considering the adoption of an LLM, it is necessary to carry out a technology assessment in order to find the most appropriate set-up given the business' need.

The efficiency of LLMs in operational environments can be significantly enhanced by integrating advanced computing techniques aimed at reducing latency. Parallel processing through data and model parallelism allows simultaneous processing across multiple units, crucial for handling large models and data batches in real-time. Techniques like caching and pre-fetching enhance responsiveness by storing results for repeated use and anticipatively loading data, respectively, which is especially beneficial in services like streaming or voice assistants. Adaptive computation adjusts the computational effort based on query complexity, optimizing processing time and resource use. Asynchronous processing and network optimization improve latency by allowing non-blocking task execution and enhancing data transmission speeds, respectively. Hardware acceleration through devices like TPUs can expedite inference times, while load balancing through smart routing and dynamic resource allocation ensures optimal server utilization and resource distribution.

Collectively, these strategies significantly reduce latency, enhancing the real-time effectiveness of LLMs across various applications including AI systems, translation services, and predictive analytics, making them more responsive, efficient, and scalable in everyday use. The selection of the appropriate model size and architecture, together with the optimization of the inference process, is essential for the effective use of LLMs in the context of NLA. The optimal setup will vary depending on the specific accuracy and response time requirements of each application. Nevertheless, it is worth considering the following: *model size*, *architecture*, *deployment* and *optimization*. Each of these elements will be presented below.

2.3.1   Model Size

Model sizes vary widely, each offering distinct advantages and challenges. Smaller models with 7B to 9B parameters are quicker and require less memory and processing, making them suitable for low-latency applications and simpler hardware setups; however, they might lack deeper language processing capabilities. Medium-sized models, ranging from 15B to 27B parameters, balance computational demands with performance, enabling them to handle more complex queries and provide detailed insights, which are feasible for many enterprise applications. Large models, such as those with 60B to 80B parameters, offer extensive contextual understanding and nuanced text generation, though they require significant computational power and often experience latency unless supported by specialized hardware. Models exceeding 150 billion parameters demand intense computational resources, making them unsuitable for local deployment and are typically operated through cloud-based platforms or specialized hardware to manage their operational demands effectively.

2.3.2   Architecture

The architecture of large language models (LLMs) like Transformers directly impacts their operational efficiency, particularly in terms of latency. Traditional Transformers [26] scale quadratically with input size, which can significantly increase latency, particularly in real-time applications. Innovations such as the Performer [27] and Linformer [28] mitigate this by linearizing or reducing the computation of the attention mechanism, thus enhancing speed and efficiency for handling longer sequences or larger datasets. Sparse Transformers [29] and the Reformer [30] introduce sparse and locality-sensitive hashing attention mechanisms, respectively, which focus computational efforts on critical data segments to decrease latency effectively. Hybrid models like MobileBERT [31] incorporate elements from various architectures to optimize performance across different tasks while maintaining reduced latency, making them suitable for environments with stringent speed requirements like mobile devices. These architectural advancements collectively aim to address the fundamental challenge of reducing latency in LLM operations.

2.3.3 Deployment Solution

Deployment solutions for model inference can either be on-device or cloud-based, each with distinct advantages and challenges concerning latency. On-device processing offers reduced latency as data doesn't need to travel over networks, making it ideal for applications requiring quick responses, like mobile translation or augmented reality. However, the computational power of local devices is limited, often requiring models to be simplified. Conversely, cloud-based processing utilizes powerful server capabilities, enabling the use of larger, more complex models suitable for demanding tasks like deep semantic analysis. The main disadvantage of cloud deployment is increased latency from data transmission, though advancements in network technologies such as 5G can mitigate this issue. Optimization techniques like quantization, pruning, and distillation are critical across both platforms, helping to reduce model size and speed up inference without significantly compromising performance, essential for latency-sensitive environments.

2.3.4 Optimization Techniques

Optimization techniques are crucial for deploying large language models (LLMs) effectively, particularly in environments where both performance and operational efficiency are priorities. Techniques such as quantization [32] reduce precision to lighten computational loads and memory demands, enhancing model responsiveness, especially in on-device contexts. Pruning [33] removes less impactful parameters, streamlining models for quicker inference and lower operational costs. Model distillation [34] transfers knowledge from a large model to a smaller, more manageable one, maintaining performance with reduced computational needs, ideal for real-time applications. Sparse training [35] and low-rank factorization [36] minimize unnecessary computations by promoting inherent sparsity and decomposing weight matrices into simpler forms, respectively, which benefits environments with limited hardware capabilities. Layer and channel skipping [37], and frequency domain method [38] adjust computational depth dynamically during inference, reducing processing time and power consumption crucial for real-time and resource-limited applications. These techniques collectively enable the efficient deployment of LLMs across various platforms, balancing latency, power, and accuracy to meet diverse application needs.

This chapter has presented the central concepts and the possible role of Natural Language Analytics (NLA) in contemporary decision-making processes, emphasizing the transformative potential that language models bring into the scenario of business intelligence. By dissecting the nuanced relationships within language, NLA provides a framework for extracting soft data, which are qualitative, subjective insights that are often overlooked by traditional analytics. The introduction of Key Soft Indicators (KSIs) formalize a new dimension of indicators, based on a more abstract metrics, expanding the comprehensive understanding of organizational dynamics. Through advanced computational techniques, the use of local LLMs, and the integration of big data, NLA can operate data processing with speed, efficiency and reliability, enriching the quality of decision-making. As these models continue to be refined, it is expected a progressive improvement on the applications, helping further with the balance between computational power and the delivery of timely, actionable insights, which is a critical focus of these sort of solutions. Because business applications handle sensitive data, they should not run in a cloud environment. To facilitate use on a local, business-level machine, a framework such as *Ollama*, described in the next chapter, is beneficial.

References

1. H. Chen, R. H. L. Chiang, and V. C. Storey. Business intelligence and analytics: From big data to big impact. *MIS quarterly*, pages 1165–1188, 2012.
2. T. H. Davenport. Analytics 3.0. *Harvard business review*, 91(12):64–72, 2013.
3. S. Chatterjee, R. Chaudhuri, S. Kamble, S. Gupta, and U. Sivarajah. Adoption of artificial intelligence and cutting-edge technologies for production system sustainability: a moderator-mediation analysis. *Information Systems Frontiers*, 25(5):1779–1794, 2023.
4. U. Dinesh Kumar. *Business analytics: The science of data-driven decision making*. Wiley, 2017.
5. S. F. Wamba, S. Akter, A. Edwards, G. Chopin, and D. Gnanzou. How 'big data' can make big impact: Findings from a systematic review and a longitudinal case study. *International journal of production economics*, 165:234–246, 2015.
6. E. D. Zamani, C. Smyth, S. Gupta, and D. Dennehy. Artificial intelligence and big data analytics for supply chain resilience: a systematic literature review. *Annals of Operations Research*, 327(2):605–632, 2023.
7. B. D. Langhe and S. Puntoni. *Decision-Driven Analytics: Leveraging Human Intelligence to Unlock the Power of Data*. University of Pennsylvania Press, 2024.
8. T. Davenport and J. Harris. *Competing on Analytics: The New Science of Winning*. Harvard Business Press, 2017.
9. K.E. Kendall and J.E. Kendall. *Systems Analysis and Design*. Pearson Prentice Hall, 2011.
10. M. J. Liberatore and W. Luo. The analytics movement: Implications for operations research. *Interfaces*, 40(4):313–324, 2010.
11. V. Sharma, J. Poulose, and C. Maheshkar. Analytics enabled decision making "tracing the journey from data to decisions". In *Analytics Enabled Decision Making*, pages 1–22. Springer, 2023.
12. G. Barros. Racionalidade e organizações: um estudo sobre comportamento econômico na obra de herbert a. simon. Master's thesis, Universidade de São Paulo, 2016.
13. T. L. Saaty. Decision making with the analytic hierarchy process. *International journal of services sciences*, 1(1):83–98, 2008.
14. K. R. R. Coutinho. A psicologia da gestalt: aplicabilidade a prática pedagógica da educação de jovens e adultos. 2008.
15. W. Kohler. *The task of Gestalt psychology*. Princeton University Press, 2015.
16. H. W.J. Rittel and M. M. Webber. Dilemmas in a general theory of planning. *Policy sciences*, 4(2):155–169, 1973.
17. M. Minelli, M. Chambers, and A. Dhiraj. *Big data, big analytics: emerging business intelligence and analytic trends for today's businesses*, volume 578. John Wiley & Sons, 2013.
18. M. Y. Chu. *Blissful Data*. Amacom, 2004.

19. Matthew J. Traxler. *Introduction to psycholinguistics: Understanding language science*. Wiley-Blackwell, 2011.
20. K. McRae and M. F. Joanisse. The cambridge handbook of psycholinguistics.
21. M. Yang and J. Liang. The handbook of psycholinguistics. *Journal of Linguistics*, 54(2):429–434, 2018.
22. R. Hackathorn. The bi watch real-time to real-value. *DM review*, 14:24–29, 2004.
23. D. Jurafsky and J. H. Martin. *Speech and Language Processing*. draft (https://web.stanford.edu/~jurafsky/slp3/), third edition, 2023.
24. P. Fu, Z. Lin, F. Yuan, W. Wang, and D. Meng. Learning sentiment-specific word embedding via global sentiment representation. In *Proceedings of the AAAI Conference on Artificial Intelligence*, volume 32, 2018.
25. D. Apgar. *Risk intelligence: Learning to manage what we don't know*. Harvard Business Press, 2006.
26. A. Vaswani, N. Shazeer, N. Parmar, J. Uszkoreit, L. Jones, Aidan N. Gomez, Ł. Kaiser, and I. Polosukhin. Attention is all you need. *Advances in neural information processing systems*, 30, 2017.
27. K. Choromanski, V. Likhoshcrstov, D. Dohan, X. Song, A. Gane, T. Sarlos, P. Hawkins, J. Davis, A. Mohiuddin, L. Kaiser, D. Belanger, L. Colwell, and A. Weller. Rethinking attention with performers. arXiv preprint arXiv:2009.14794, 2020.
28. S. Wang, B. Z. Li, M. Khabsa, H. Fang, and H. Ma. Linformer: Self-attention with linear complexity. arXiv preprint arXiv:2006.04768, 2020.
29. R. Child, S. Gray, A. Radford, and I. Sutskever. Generating long sequences with sparse transformers. arXiv preprint arXiv:1904.10509, 2019.
30. N. Kitaev, Ł. Kaiser, and A. Levskaya. Reformer: The efficient transformer. arXiv preprint arXiv:2001.04451, 2020.
31. Z. Sun, H. Yu, X. Song, R. Liu, Y. Yang, and D. Zhou. Mobilebert: a compact task-agnostic bert for resource-limited devices. arXiv preprint arXiv:2004.02984, 2020.
32. B. Jacob, S. Kligys, B. Chen, M. Zhu, M. Tang, A. Howard, H. Adam, and D. Kalenichenko. Quantization and training of neural networks for efficient integer-arithmetic-only inference. In *Proceedings of the IEEE conference on computer vision and pattern recognition*, pages 2704–2713, 2018.
33. S. Han, H. Mao, and W. J. Dally. Deep compression: Compressing deep neural networks with pruning, trained quantization and huffman coding. arXiv preprint arXiv:1510.00149, 2015.
34. G. Hinton, O. Vinyals, and J. Dean. Distilling the knowledge in a neural network. arXiv preprint arXiv:1503.02531, 2015.
35. D. C. Mocanu, E. Mocanu, P. Stone, P. H. Nguyen, M. Gibescu, and A. Liotta. Scalable training of artificial neural networks with adaptive sparse connectivity inspired by network science. *Nature communications*, 9(1):2383, 2018.
36. M. Jaderberg, A. Vedaldi, and A. Zisserman. Speeding up convolutional neural networks with low rank expansions. arXiv preprint arXiv:1405.3866, 2014.
37. X. Wang, F. Yu, Z. Dou, T. Darrell, and J. E. Gonzalez. Skipnet: Learning dynamic routing in convolutional networks. In *Proceedings of the European conference on computer vision (ECCV)*, pages 409–424, 2018.
38. Y. Wang, C. Xu, S. You, D. Tao, and C. Xu. Cnnpack: Packing convolutional neural networks in the frequency domain. *Advances in neural information processing systems*, 29, 2016.

Chapter 3
Using Ollama

Be like the llama: climb mountains, hold your head high and don't spit into the wind.

— Anonymous

Abstract Ollama is a tool designed to facilitate the deployment and operation of Large Language Models (LLMs) for various language analytics tasks. This chapter provides a quick-start guide for Ollama, detailing the steps to download and start using the tool on a local machine. This includes the navigation of Ollama's model library and selection of models, the use of Ollama in a command shell environment, the setup of models through a `modelfile`, and its integration with Python (enabling developers to incorporate LLM functionality into Python-based projects). Ollama is presented as a solution that not only enhances the use of LLMs but also makes them more adaptable and easier to use in various contexts.

Ollama is an easy-to-use framework for running LLMs locally on either CPU or GPU that can be used for supporting NLA, as show in Prompt 2.1. Perhaps its main advantage is that it provides a user-friendly interface for downloading, setting up, integrating, running and sharing open-source LLMs. The same task can be done with Huggingface (https://huggingface.co/), but with a technology overhead that is transparent in Ollama. In addition, to run LLM on a business-level computer, it is necessary to use backends for efficient management of the CPU (and low-scale GPU), such as the `llama.cpp` (https://github.com/ggerganov/llama.cpp), another technology overhead that is circumscribed by Ollama. In short, Ollama encompasses the underlying technology for running LLMs locally, and provides endpoints in the form of an API, as well from direct access by programming languages, such as a Python module. For an in-depth presentation, check out Ollama's links at:

Homepage: https://ollama.com/, for the binaries, news and model catalog;
GitHub: https://github.com/ollama/ollama, to source, document and learn more;
PyPi: https://pypi.org/project/ollama/, for the Python module.

F. S. Marcondes et al., *Natural Language Analytics with Generative Large-Language Models*, SpringerBriefs in Computer Science,
https://doi.org/10.1007/978-3-031-76631-2_3

Table 3.1 Quantization comparison on Llama2 [2] on *perplexity* and *file size* (lower is better for both metrics). Column F16 is the base model with 16-bit floats, columns starting with Q4 and Q5 are 4-bit and 5-bit quantisations respectively. The next symbol indicates the rounding method, _0 and _1 are two different types of uniform quantization, and _K is an approach that aims to optimise memory usage by using different bit widths. Finally, the symbols _S, _M and _L refer to the size of the blocks used for quantization, from small to large. Roughly speaking, quantizing with _0 or _S focuses on speed, while _1 or _L focuses on performance [3]

Llama2	Measure	F16	Q4_0	Q4_1	Q5_0	Q5_1	Q4_K_S	Q4_K_M	Q5_K_S	Q5_K_M
7B	perplexity	5.9066	6.1565	6.0912	5.9862	5.9481	6.0215	5.9601	5.9419	5.9208
7B	file size	13.0G	3.5G	3.9G	4.3G	4.7G	3.6G	3.8G	4.3G	4.5G
13B	perplexity	5.2543	5.3860	5.3608	5.2856	5.2706	5.3404	5.3002	5.2785	5.2638
13B	file size	25.0G	6.8G	7.6G	8.3G	9.1G	6.8G	7.3G	8.4G	8.6G

An important development for running LLMs locally, which is also quite transparent in Ollama, is *quantization* (for a full featured explanation, refer to [1]). In short, it is the reduction of the floating point precision used in the model (usually from 16 to 4 bits long). As expected, the file size shrinks considerably, but the performance of the model (often measured by *perplexity*, i.e. how well a probability model predicts a sample) also deteriorates. The point is that the performance does not deteriorate at the same rate as the model size decreases, which results in a reasonable cost-benefit ratio, for a reference see the Table 3.1. It is worth noting in this table that quantization can be the difference between a model with reasonable performance requiring special hardware and running on most average computers.

Note that the base models provided by Ollama on its homepage are often quantized to Q4_0, but in its catalog the community provides most models with all types of quantization. Note also that although Ollama has its own catalog, it is probably designed to provide a straightforward experience in the sense that almost any model in Huggingface will also run in Ollama after a quick setup (which still creates an overhead for newcomers). It should be noted that Ollama is developing at a rapid pace, with new features and technology support being added all the time. Instead of providing an extensive presentation of Ollama, it is intended in this chapter to introduce its basic principles by exploring some of its key features, allowing the reader to easily operate other features and follow further developments.

The accessibility provided by Ollama means that high-end GPUs or specialized hardware are often not required, significantly lowering the cost and barrier to entry for implementing LLMs as a widespread tool. As an added bonus, little to no programming knowledge is required (apart from a few command lines, depending on the task). Overall, `llama.cpp` and Huggingface perhaps offer more control over the engine. However, Ollama offers ease of use and a streamlined experience by simplifying things, making it a great choice for beginners looking for an out-of-the-box experience that can scale up to more complex systems.

3.1 Quick-Start

To get started with Ollama, visit the Ollama homepage and navigate to the download section. Then, select the respective operating system, and start downloading. The download should take only about a minute. On Linux, the installation can be undertaken in the Command Shell, see Listing 3.1. For a reference, the whole process is depicted in Fig. 3.1. For bridging with the notation used in this book, Prompt 3.1 illustrates the effect of writing "Hello!" on the last line of Fig. 3.1 (after the »>). To exist the chat, the user must type "Ctrl + D" or /bye at the conversation prompt.

Listing 3.1 Installation of Ollama on Bash

```
1  $ curl -fsSL \url{https://ollama.com/install.sh} | sh
2  $ ollama run phi3:latest
```

Prompt 3.1 phi3:latest

```
1  >>> Hello!
2  Hello! How can I assist you today?
```

A caution to be taken on selecting a model to install is to assert that it fits into the available RAM of the computer (on Linux type free -m). The size of the model in this example is of about 2Gb, Llama3 with 70B parameters requires around 40Gb whereas Qwen2 with 0.5B parameters, about 300Mb. The size of the model will certainly influence the performance of the prediction, then the need for finding a

```
$ sudo curl -fsSL https://ollama.com/install.sh | sh
[sudo] senha para    :
>>> Downloading ollama...
############################################################### 100,0%#
############################################################### 100,0%
>>> Installing ollama to /usr/local/bin...
>>> Adding ollama user to render group...
>>> Adding ollama user to video group...
>>> Adding current user to ollama group...
>>> Creating ollama systemd service...
>>> Enabling and starting ollama service...
>>> The Ollama API is now available at 127.0.0.1:11434.
>>> Install complete. Run "ollama" from the command line.
WARNING: No NVIDIA/AMD GPU detected. Ollama will run in CPU-only mode.
$ ollama run phi3:latest
pulling manifest
pulling 3e38718d00bb... 100%              2.2 GB
pulling fa8235e5b48f... 100%              1.1 KB
pulling 542b217f179c... 100%              148 B
pulling 8dde1baf1db0... 100%               78 B
pulling ed7ab7698fdd... 100%              483 B
verifying sha256 digest
writing manifest
removing any unused layers
success
>>> Send a message (/? for help)
```

Fig. 3.1 Installation of Ollama on Bash

good balance. Check the model library at Ollama's homepage for finding a suitable model to the available hardware resources. Remember that the base models provided by Ollama on its homepage are often quantized to Q4_0.

3.2 Using Ollama

Ollama works in the background, so once installed it can be used either in the command shell, via a Python interpreter, or as a web service (for stop the service on Linux use `$ systemctl stop ollama.service`). Ultimately, any Ollama operation can be performed in any mode, but in practice some tasks are better suited to one way or another. In short, for the purposes of this book, downloading and setting up a model is better performed on Bash while the use of the model for specific tasks in Python (the web-service is omitted, if needed, check Ollama's documentation).

3.2.1 Operations on Command Shell

The main command `$ ollama run MODEL` was introduced in the last section. Note that the command `$ ollama pull MODEL` is called implicitly when a model has not yet been downloaded, so the command `pull` is called when it is only necessary to download the model without running it. Depending on the use case, it is useful to pass the prompt as a parameter to the `run` command, as shown in the listing 3.2 (in this case no context window is involved).

Listing 3.2 Passing a prompt using CLI
```
1  $ ollama run phi3:latest "Hello!"
2  Hello there! How can I help you today?
```

It is worth mentioning that Ollama also support multi-modal models. An instance of a prompt with an image file queried to `Llava` is presented in Prompt 3.2.

Prompt 3.2 Multimodal prompt example in Ollama
```
1  >>> What is in /Images/smile.png?
2  The image contains a simple graphic of a smiling face. The face
     ↪ consists of basic shapes and lines that give it an
     ↪ abstract and minimalist appearance. There are no other
     ↪ objects or text within the image, which focuses solely on
     ↪ the depiction of the smile.
```

The way to find the model to `pull` and `run` is by evaluating its *modelcard*, for reference an example is shown in Fig. 3.2. In short, the combo box lists all the model variations in the catalog to be selected, the whole command to run the selected model is presented in the text box next to it. Below the combo box there is a time stamp showing the last update. Below the timestamp is the model description with the base

phi3

Phi-3 is a family of lightweight 3B (Mini) and 14B (Medium)
state-of-the-art open models by Microsoft.

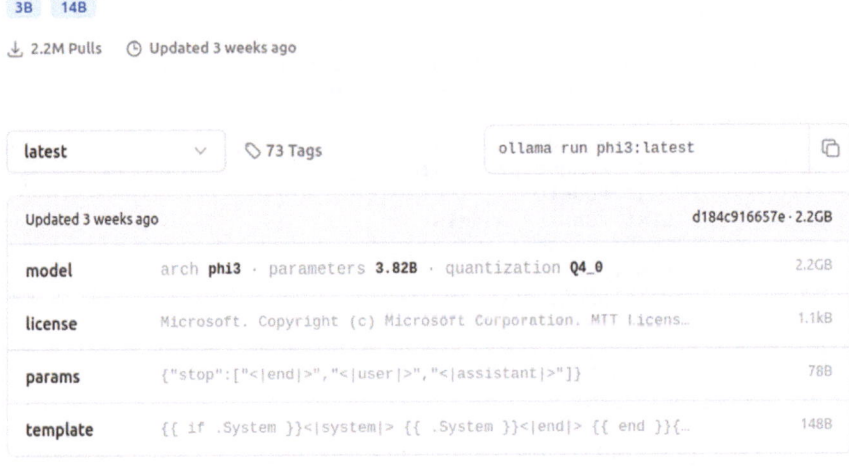

Fig. 3.2 Ollama model card for Phi3 (https://ollama.com/library/phi3:latest)

name of the model, its number of parameters, the quantization method used and the
size to be taken into account when running it. The next line is the first few words
of the license agreement that is implicitly accepted when using the model. Note that
both the *model* and *license* words are links that can be followed for details. The other
lines will be presented latter in this chapter.

A list of all bash commands is shown in the Fig. 3.3. For everyday use, perhaps the
most useful commands in addition to those already mentioned are `$ ollama list`
to list the local models and `$ ollama rm MODEL` to remove some of them. For trou-
bleshooting, it is advisable to keep track of Ollama versions through regular updates.
Selecting the perfect model can be challenging due to overlapping functionalities
and the unique nature of each task. While benchmarks offer insights into model performance, they might not cover all possible questions or use cases. The best way
to identify the most suitable model is to test each one personally and observe their
performance in specific scenarios.

3.2.2 Customized Model Setup

Another fundamental feature of Ollama is the possibility of setting-up a given model,
suiting it for specific needs. This is usually performed through a `modelfile` that is
passed as parameter to the `create` command (refer to Fig. 3.3). A `modelfile`

```
$ ollama
Usage:
  ollama [flags]
  ollama [command]

Available Commands:
  serve       Start ollama
  create      Create a model from a Modelfile
  show        Show information for a model
  run         Run a model
  pull        Pull a model from a registry
  push        Push a model to a registry
  list        List models
  ps          List running models
  cp          Copy a model
  rm          Remove a model
  help        Help about any command

Flags:
  -h, --help        help for ollama
  -v, --version     Show version information

Use "ollama [command] --help" for more information about a command.
```

Fig. 3.3 Ollama Bash Commands

Table 3.2 Instructions of Ollama's `modelfile`

Instruction	Description
FROM (required)	Defines the base model to use
SYSTEM	Specifies the system message that will be set in the template
MESSAGE	Specify message history
PARAMETER	Sets the parameters for how Ollama will run the model
TEMPLATE	The full prompt template to be sent to the model
ADAPTER	Defines the (Q)LoRA adapters to apply to the model
LICENSE	Specifies the legal license

is a text file describing the setup of a given model with the form INSTRUCTION
<arguments>, a list with possible instructions is presented in Table 3.2.

3.2.2.1 Instruction FROM

The FROM instruction is the only required on any modelfile. It is used both to
creating a custom model (common use) and to import, and quantize, a model from

Huggingface (rarely necessary for the average user). However, since the import use provide a minimal and straightforward example, let's consider it briefly. In short, it is necessary to download the GGUF (GPT-Generated Unified Format) file of the model and produce a text document as simple as the one presented in Prompt 3.3 (for setup it would be the case of calling the model, e.g. FROM phi3:latest).

Prompt 3.3 Minimal modelfile

```
1  FROM /path/to/file.gguf
```

Once the modelfile is ready, the next step is to run it from bash with the create command. Since importing a model into Ollama would imply running it locally, and with limited hardware resources it would be desirable to quantize the model on import, the command that covers these two needs is in the Listing 3.3. The same command, but applied to the creation of a custom model, is shown in the Listing 3.4, note that the argument MY_MODEL_NAME is any name that the user wants to call this custom model.

Listing 3.3 Setup of a model in Ollama using a modelfile

```
1  $ ollama create -q Q4_K_M MODELFILE_NAME.txt
```

Listing 3.4 Setup of a model in Ollama using a modelfile

```
1  $ ollama create MY_MODEL_NAME -f MODELFILE_NAME.txt
```

3.2.2.2 Instruction SYSTEM

Perhaps the most important statement to consider when setting up a model is the System Prompt. Roughly, it can be thought of as part of the prompt that would be repeated on each call. As a simple example, consider an LLM that should be suitable for translation, instead of writing a prompt like

Translate the sentence from Portuguese to English: *Boa é a vida, mas melhor é o vinho.*

it is possible to set up the System Prompt with Translate the sentence from Portuguese to English using a formal tone so that only the text to be translated is included in the User Prompt. For reference, prompt 3.4 provides the full modelfile used in this example, then use the command in listing 3.4 as $ ollama create translationAssitant -f modelfile.txt, then run the newly created model with $ ollama run translationAssitant, enabling the model to be queried as in prompt 3.5. This is the structure used in Sect. 4, where Prompt 4.5 is the System Prompt and Prompt 4.6, 4.7,4.8 are User Prompts.

Prompt 3.4 Instance of a modelfile defining System Prompt

```
1  FROM phi3:latest
2
3  SYSTEM """Translate the sentence from Portuguese to English"""
```

Prompt 3.5 Prompting the custom model defined by the `modelfile` in Prompt 3.4

```
1  >>> Boa \'{e} a vida, mas melhor \'{e} o vinho.
2  The translation of this phrase into English would be: 'Life is
   ↪ good, but better still is wine.'
```

3.2.2.3   Instruction MESSAGE

Note that although the prompt 3.5 has provided the correct answer, it may not be in the most appropriate format for further processing. To improve the output format, it is possible to use few-shot learning (see Ch. 4) to guide the generation. In Ollama, this can be done by including interaction examples in the `modelfile` using the command MESSAGE <role> <message>, which will be inserted as previous prompts when Ollama starts the model (this does not work when prompting by parameter, as in Listing 3.2). The <role> can be either user or assistant, and the <message> are expected input and output texts. For example, Prompt 3.6 extends Prompt 3.4 with two examples, and Prompt 3.7 presents an instance for the extended custom model.

Prompt 3.6 Instance of a `modelfile` defining System Prompt

```
1  FROM phi3:latest
2
3  SYSTEM """Translate the sentence from Portuguese to English"""
4
5  MESSAGE user Tenho em mim todos os sonhos do mundo.
6  MESSAGE assistant I have within me all the dreams in the world.
7
8  MESSAGE user N\~{a}o tenhamos pressa, mas n\~{a}o percamos tempo.
9  MESSAGE assistant Let's not rush, but let's not waste time.
```

Prompt 3.7 Prompting the custom model defined by the `modelfile` in Prompt 3.6

```
1  >>> Tenho em mim todos os sonhos do mundo.
2  I have within me all the dreams in the world.
3
4  >>> N\~{a}o tenhamos pressa, mas n\~{a}o percamos tempo.
5  Let's not rush, but let's not waste time.
6
7  >>> Boa \'{e} a vida, mas melhor \'{e} o vinho.
8  Life is good, and better still is wine.
```

3.2.2.4   Instruction PARAMETER

The PARAMETER instruction allows the setting of (currently) fourteen parameters that define how the model is expected to behave. It is not possible in this short book to describe and exemplify each of these parameters, so the reader is invited to read the Ollama and `llama.cpp` documentation and experiment with different setups. The usage format, following the others, is PARAMETER

<parametervalue> contained in the modelfile. In short, it is possible to divide these parameters into two groups, one related to the *number of tokens* to be considered or produced, and the other related to the *selection of tokens* to be output by the model.

For the first group, the parameter num_ctx (default 2048) define the size of the context window when the next token is generated. Assuming a context window of size 10 with a prompt size of 20 tokens, only the last 10 tokens will be considered in the generation; moreover, once the size limit is reached, the context window becomes a sliding window in the sense that it "forgets" tokens in the head as news tokens are produced in the tail. It is therefore necessary to estimate the size of both the prompt and the response, and also whether chaining is expected for each use case. On the other hand, the parameter num_predict (default 128, special values are -1 for infinite generation and -2 for matching the context window) determines the maximum length of the response. Again, it will vary according to the use case. The parameter stop is used for telling the model to return when reaching a specific pattern. A known problem on LLMs is that they eventually start to produce continuously the same token, for preventing such situations, parameter repeat_last_n (default 64, special values are 0 for disabling the feature and -1 for matching the context window) defines how far the model look back in the context window for avoiding repetitions; then parameter repeat_penalty (default 1.1) determine the penalty for a repetition.

For the second group, it is worth recalling that before inserting a token in the output, the LLM calculates the probability that a word in the vocabulary will be selected as the next token. In this sense, the parameter top_k (default 40) tells the model to select one of the 'k' tokens with the highest probability, alternatively top_p (default 0.9) tells the model to select a token if its probability is higher than 'p', yet another alternative is min_p (default 0.0) which defines the minimum probability for a token to be considered. After finding a suitable set of tokens, the model randomly selects one to include in the output, so the parameter seed (default 0) sets the seed to be used by the generator, defining a non-zero value defines the seed making the model generate the same text for the same prompt. Parameter temperature (default 0.8) changes the probability distribution generated for the vocabulary by modifying a parameter of the softmax. For a reference, in the limit, on reaching zero the vocabulary becomes a one-hot vector, thus making the model to get always the same next token.

In addition to these parameters, still in the second group, there are also alternative methods that can be enabled. The tfs_z parameter (default 1, so disabled) also aims to avoid less likely tokens, but by following the rate at which probabilities decrease and truncating the tail after the 'z' threshold. The parameter mirostat (default 0, so disabled) tells the model to use the Mirostat algorithm (presumably [4]), which also aims to select tokens, but by considering perplexity instead of probability. Then mirostat_eta (default 0.1) and mirostat_tau (default 5.0) are parameter of the Mirostat, the first define the learning rate and the second the target entropy.

3.2.2.5 Instruction TEMPLATE

Training an LLM involves predicting collocations, guided by perplexity reduction. In the vocabulary of an LLM there is a special token (e.g. [EOS]) to indicate when the generation has been predicted. Using a model trained to generate sequences as a chat device requires further tuning, instruction tuning (train the model to follow instructions) and alignment tuning (the RLHF, train the model to match human expectations), are two common instances. At this stage, it is necessary to include additional special tokens to indicate that the assistant is not expected to continue with the user input, but to continue with the generation in the role of the assistant. These special tokens define the *template* of the model.

There is currently no consensus on the format, but ChatML (or variations of it) is becoming increasingly popular. For huggingface models, information about the template used is usually found in the configuration files used for training. Ollama extends llama.cpp by treating templates as Go language templates (see https:// pkg.go.dev/text/template), which opens space, yet to be explored, for sophisticated prompt customization. The template of a model is represented in the model card (see Fig. 3.2) together with special parameters such as the *stop tokens*. The template can be changed by defining it in the modelfile with the form TEMPLATE """<template description>""" (similar to the definition of the SYSTEM command). Note that when using templates, it is necessary to take into account the template on which the model was trained.

3.2.2.6 Instruction ADAPTER

The adapters refer to either Low-Rank Adaptation (LoRA) or Quantized LoRA (QLoRA). LoRA is an approach to fine-tuning LLMs that avoids training the entire model. In a simplified form *cf.* [5], the pre-trained model (W_0) is kept as is, but another, lower-rank model (ΔW) is trained on the target data. The adapter ΔW is the result of multiplying two matrices A and B, taking the dimension of W_0 as d, where the dimensions are $r \times d$ and $d \times r$ respectively (r is the rank). The adapter is stored separately from the base model, and both compute the values for the input individually, adding the resulting matrices to pass to the next level. In this sense, an adapter can be trained by the end user or reused by a third party. It is then simple, the modefile must contain the path to both the base model (the FROM instruction) and the adapter of the form Adapter <path to adapter>. A current limitation of Ollama is that the adapter must be in a GGUF file format, which may require the use of a method such as the convert_lora_to_gguf from the llama.cpp (see the llama.cpp documentation at https://github.com/ggerganov/llama.cpp) and a skeleton in the listing 3.5

Listing 3.5 Skeleton of converting a Huggingface adapter into GGML

```
1  $ huggingface-cli download <adapter>
2  $ python3 llama.cpp/convert\_lora\_to\_gguf.py <path to adapter>
```

3.2.2.7 Instruction LICENSE

It is a text description of the conditions for the use of the customized model, also part of the model card (see Fig. 3.2).

3.2.3 Operations on Python

The use of Ollama in Python is quite straightforward. First, install the module, in `pip` use `$ pip install ollama`. Then import that module on the script with `import ollama`. As mentioned, anything that can be done in Bash is also possible to be done in Python, however, features such as pulling and customizing the model are, perhaps, better suited in CLI. Nevertheless, a parallel with all the commands can be drawn, then not covered again in this section, a detailed explanation can be found at the module repository on GitHub (https://github.com/ollama/ollama-python). The advantage of using Python to interact with an LLM is that certain processes required for NLA, such as extracting features from a dataset, can be automated, allowing language computations to be performed at scale and in integration with other NLP and vizualiation modules e.g. `pandas`, `SpaCy`, `matplotlib`, *etc*. The general form of a Python script using the Ollama module is presented in Listing 3.6.

Listing 3.6 Example of using Ollama in Python

```
1  import ollama
2  Prompt = [
3    {'role': 'system', 'content': 'Translate the sentence from
         ↪ Portuguese to English'},
4
5    {'role': 'user', 'content': 'Tenho em mim todos os sonhos do
         ↪ mundo.'},
6    {'role': 'assistant', 'content': 'I have within me all the
         ↪ dreams in the world.'},
7
8    {'role': 'user', 'content': 'N\~{a}o tenhamos pressa, mas n\~{a
         ↪ }o percamos tempo.'},
9    {'role': 'assistant', 'content': 'Let\'s not rush, but let\'s
         ↪ not waste time.'},
10
11   {'role': 'user', 'content': 'Boa \'{e} a vida, mas melhor \'{e}
         ↪   o vinho.'},
12        ]
13
14 response = ollama.chat(model='tinyllama:latest', messages=Prompt)
         ↪
15 print(response['message']['content'])
```

As you can see from the listing 3.6, the `prompt` is a list of dictionaries with a `role`, which can be either `system`, `user` or `assistant`. This reproduces the behavior of the SYSTEM and MESSAGE commands above. On the other hand, including only the

```
>>> ollama.
ollama.AsyncClient(      ollama.Options(            ollama.copy(            ollama.list()
ollama.ChatResponse(     ollama.ProgressResponse(   ollama.create(          ollama.pull(
ollama.Client(           ollama.RequestError(       ollama.delete(          ollama.push(
ollama.GenerateResponse( ollama.ResponseError(      ollama.embeddings(      ollama.show(
ollama.Message(          ollama.chat(               ollama.generate(
```

Fig. 3.4 The Ollama module's operation list

last element of the list has the same effect as querying a prompt when running a model in the CLI. The operations list of the Ollama module is shown in Fig. 3.4 and the signature for three operations of interest are shown below:

- `ollama.embeddings`(model, prompt, options, keep_alive)
- `ollama.chat`(model, messages, stream, format, options, keep_alive)
- `ollama.generate`(model, prompt, system, template, context, stream, raw, format, images, options, keep_alive)

The example presented in listing 3.6 uses the `chat` operation in line 14. Note that `generate` could also be used in the same context. A rule of thumb would be to use `generate` when no context is needed and `chat` when using strategies such as few-shot learning. It would also be good practice for the general case not to include the system prompt (or perform other model-related operations) in the way that is done in this example, but by defining a custom model. The `embeddings` operation returns the embedding of the prompt produced by the model instead of the final tokens. This is necessary for Retrieval Augmented Generation (RAG) [6], where the embeddings of the input are used to find similar embeddings of other texts to enrich the response generation (search for embedding models in Ollama's catalog). Note, from the operations list, that it is also possible to interact with Ollama asynchronously, yet a discussion beyond the scope of this book. For a reference, the `response` returned in line 14 by the `ollama.chat` method is presented in Listing 3.7.

Listing 3.7 Response of the `ollama.chat` method.

```
1  {'model': 'tinyllama:latest', 'created_at': '2024-07-31T10
   ↪  :00:29.313777377Z', 'message': {'role': 'assistant', '
   ↪  content': 'Good life, better wine? Yes please!'}, '
   ↪  done_reason': 'stop', 'done': True, 'total_duration':
   ↪  3779463714, 'load_duration': 1284036655, '
   ↪  prompt_eval_count': 163, 'prompt_eval_duration':
   ↪  1811834000, 'eval_count': 10, 'eval_duration': 414729000}
```

3.3 Ollama Ecosystem

It is not possible to describe the whole Ollama ecosystem in detail in a book, but it receives a lot of community support and developer-oriented effort. Such support is manifested in the generation of all kinds of models and customizations, the supportive mutual help of its user base, as well as the expanding landscape of tools based on Ollama.

For a quick overview, community-developed tools have democratized access to Ollama's capabilities, Open WebUI (https://github.com/open-webui/open-webui) is a meaningful instance. These user-friendly interfaces have lowered the barrier to entry, enabling individuals without extensive technical expertise to experiment with and benefit from advanced language models. By providing intuitive platforms for interaction, these tools have contributed significantly to the growth and adoption of Ollama.

Developer-centric tools like LangChain (https://github.com/langchain-ai/langchain), on the other hand, have transformed Ollama into a versatile building block for complex AI applications. These libraries offer a rich toolkit for constructing sophisticated language models, empowering developers to create innovative solutions tailored to specific needs. From chatbots to document summarizers, the possibilities are vast.

Moreover, the integration of Ollama with other technologies and platforms has further expanded its utility. For instance, its compatibility with various programming languages and frameworks allows for seamless integration into existing applications. This interoperability has fostered a thriving ecosystem of complementary tools and services, enhancing Ollama's overall value proposition.

In a nutshell, *Ollama* is an easy-to-use platform for deploying and managing LLMs locally. By providing a set of convenient features that allow a smooth use and setup of LLMs, from quantization to retrieving embeddings, and finally to function calling for further integration into a full-featured program or data pipeline. Ollama is under rapid development, and this last feature is an example of a recent integration that has not been discussed in this chapter. With the framework for running LLMs locally in place, it is time to discuss prompt engineering as a way of refining LLM interactions and ensuring they are tailored to specific needs, discussed in the next chapter.

References

1. M. Grootendorst. A visual guide to quantization. https://towardsdatascience.com/a-visual-guide-to-quantization-930ebcd9be94, 2024.
2. G. Gerganov. The llama.cpp documentation. https://github.com/ggerganov/llama.cpp/blob/master/examples/quantize/README.md, 2024.
3. A. Shatokhin. Llms quantization naming explained. https://andreshat.medium.com/llm-quantization-naming-explained-bedde33f7192, 2024.
4. S. Basu, G. S. Ramachandran, N. Keskar, and L. R. Varshney. Mirostat: A neural text decoding algorithm that directly controls perplexity, 2021.
5. E. J. Hu, Y. Shen, P. Wallis, Z. Allen-Zhu, Y. Li, S. Wang, L. Wang, and W. Chen. Lora: Low-rank adaptation of large language models, 2021.
6. P. Lewis, E. Perez, A. Piktus, F. Petroni, V. Karpukhin, N. Goyal, H. Küttler, M. Lewis, W. Yih, T. Rocktäschel, et al. Retrieval-augmented generation for knowledge-intensive nlp tasks. *Advances in Neural Information Processing Systems*, 33:9459–9474, 2020.

Chapter 4
Generative Prompt Engineering

I recently told my daughter, a college student: If you want to pursue a career in engineering, you should focus on learning philosophy in addition to traditional engineering coursework. Why? Because it will improve your code.

— Marco Argenti, 2024

Abstract This chapter delves into the methodologies of prompt engineering to enhance the functionality and precision of Large Language Models (LLMs). It begins by highlighting the role of prompt engineering in aligning LLM outputs with specific user needs and intentions. Then categorizes prompting techniques into several types, each designed to improve how LLMs understand and respond to tasks: Zero-shot prompting; One-shot prompting; Few-shot prompting; and Chain-of-thought prompting. Chain-of-thought guide the LLM through a logical sequence to solve tasks requiring detailed analysis. By employing semiotics-the study of signs and symbols and their use or interpretation-the prompts can be designed to navigate the model through structured reasoning paths, addressing both the syntactic and semantic layers of language processing. This perspective helps in crafting prompts that act as mediating agents between the user and the LLM, aligning the machine's outputs with human semantic intentions and enhancing the interaction quality.

The ability to create descriptions of mental models aiming at the problems to be tackled is a critical skill in the context of AI. In this sense, the application of philosophy in the context of prompt engineering (perhaps, still, prompt craftsmanship) presents a possible path for guiding the underlying biases (or "tendencies") of LLMs. Thus,

F. S. Marcondes et al., *Natural Language Analytics with Generative Large-Language Models*, SpringerBriefs in Computer Science,
https://doi.org/10.1007/978-3-031-76631-2_4

it is expected that applied philosophy[1] [4–6] aids both on the inception of principles and on the effective design of prompts grounded on concepts as such.

A *physical symbol*, proposed by Simon and Newell, is a physical pattern (such as marks on paper or arrangements of bits in a computer) that is manipulated according to given rules to produce a desired effect [7, 8]. They then suggest that cognitive processes are fundamentally computational (both being a *physical symbol system*) because both processes are analogous, a strong AI view. A weaker and perhaps more accurate version of this hypothesis is that some cognitive processes, which are mostly mechanical, can also be performed by a computer. Nevertheless, it provides the necessary bridge between computer science and some philosophical notions.

Intentionality is the thoughts and interpretations of the interpretative act itself [9], i.e. a movement of how the being extends upon the world. Among several hypotheses, *physicality* suggests that intentionality is given by motricity and bodily perception [10] enabling it to be linked with physical symbol systems. In short, these systems operate by creating and manipulating symbols to represent various elements of the world, thereby allowing for the modelling of intentionality within mechanical processes. The implication is that a computer system have a *designed intentionality* [11], distilled from the human intentionality.

Generally, most of the LLMs are designed to not deliver negative outputs to a prompt, this is the *intentionality, tendency*, designed for the model. In this sense, by asking a model as such for classifying the sentiments of a phrase into `positive` or `negative` would have a bias towards a positive classification. By replacing the negative word with `less positive` help reducing the bias. For instance, if you prompt a model for levels of "emotional stability" may trigger better results than prompting it to classify in "calm vs neurotics". The positive biased model tends to avoid sensible classification as neurotics.

In the context of computational models, the notion of the connectionist paradigm offers a departure from classical theories of symbol manipulation, with inspiration from cerebral processes [12, 13]. In short, during the training phase, a neural network creates an embedding space from which the intentionality of the model emerges rather than being predefined. So it is not about *designed intentionality* but about *emergent intentionality*. Currently, there are no good ways to approach the embedding space generated by a neural network, and LLMs are regarded as "black boxes" after training [14].

An alternative paradigm that can be used for approaching the LLMs' tendencies is the semantic paradigm of communication [15] which aims to understand the interplay between language, meaning, and intentionality. Rooted in linguistic theories and

---

[1] Applied philosophy is the use of philosophical modes of thought to address practical problems. For example, in the context of this book, the weak Sapir-Whorf hypothesis (or language relativism hypothesis) suggests that the written expressions of a culture embed its moral values [1]; since an LLM models the language of a culture, it is possible to suggest that its biases would follow the culture embedded in the dataset of each language. In this sense, what is commonly referred to as the bias of the model reflects the *tendencies* of a culture as expressed by its dataset. Evidence for this possibility is provided by the "Right-wing GPT" initiative [2] and the differences in results between languages [3].

semantic networks [16], this paradigm emphasizes the representation and processing of meaning within computational systems. By focusing on the relationships between symbols and their meanings, semantic models aims to capture the nuances of human language and thought processes embedded in the communications processes.

The semantic knowledge paradigm applied to LLMs provide an understanding on how do the LLMs interpret, generate, and reason about natural language from an external perspective. This paradigm is particularly relevant in the context of LLMs, which rely on vast amounts of textual data to learn the semantic associations and contextual subtleties of language. By leveraging semantic information, these models can enhance their ability to generate coherent and contextually appropriate responses, thereby exhibiting a form of underlying intentionality through their interactions with users.

To sum it up, the operation domain of LLMs resembles their prototyped behavior, positing that LLMs can show a form of "embedded" or "tacit" intentionality, which arises from the tendencies inherent in their training sets and processing modules. This overall embedded features equips LLMs with the capacity to predict the next token through a certain bias, in the scope of semantic paradigm, and expressed in the form of words, phrases, texts, or logics. This perspective extends the concept of intentionality, treating technologies as such as active agents that also mediate and influences human actions [17].

In this context, prompt engineering is a critical field of study due to its pivotal role in shaping and guiding the behavior and capabilities of LLMs to align with user intentions and needs. By constructing prompts that act as mediating agents, prompt engineering helps with bridging the gap between the inherent biases and tendencies embedded within LLMs and the desired outcomes specified by the operator. This process involves carefully designing and refining instructions to guide LLMs in generating analysis and responses that are coherent, object centred, contextually appropriate, and aligned with user expectations.

4.1 Generative Prompting Techniques

Prompting involves providing a specific set of instructions or inputs to elicit tailored responses from an LLM [18]. This technique makes use of existing knowledge gained from training on different datasets. Since small variations in the prompt result in large differences in the output, several prompt approaches have been developed [19].

Perhaps the most common approaches[2] are currently `zero-shot`, `one-shot`, `few-shot` and `chain-of-thought` [20–22].

`Zero-shot` prompt involves providing an LLM with a task without any examples of how the response ought to be, relying solely on the pre-trained capabilities of the model. The main advantage of this approach is its simplicity, as it requires no additional data or examples to generate a response. This makes it highly versatile and quick to implement across a wide range of tasks. The downside is that the model's performance can be inconsistent and less accurate for tasks it has not trained to perform. The lack of examples can lead to misunderstandings or less relevant responses, due to the vagueness or ambiguity that a short prompt may carry, making it less reliable for highly specialized or nuanced queries.

`One-shot` prompt provides the model with a single example to guide its response to a new task. This approach adds some context and precision when compared to the zero-shot approach. But stills falls behind of `few-shot` prompting performance. By providing a single example, it helps the model understand the expected output, improving accuracy and relevance, and reducing vagueness and ambiguity. The advantage here is that it still requires minimal data for enhancing the model's performance compared to zero-shot learning. However, the limitation is that one example may not be sufficient for the model to fully grasp complex or highly varied tasks, potentially leading to less optimal responses if the provided example is not representative enough.

`Few-shot` involves providing the model with a set of examples in conjunction with the prompt. This approach significantly enhances the model's understanding and accuracy by giving it a broader context and a clearer idea of the expected output. The primary advantage of `few-shot` prompting is its ability to improve performance on complex or nuanced tasks by offering diverse examples. However, this method requires more effort in curating relevant examples. The quality and representativeness of the examples are crucial, as poor examples can lead to misunderstandings and incorrect responses, misguiding the model. It is worth mentioning that the use of this strategy often outperform fine-tuning [18].

`Chain-of-thought` prompting involves guiding the model through a series of logical steps or reasoning processes to solve complex problems, leading it to a conclusion. This method leverages the model's ability to follow step-by-step reasoning, making it particularly useful for tasks that require detailed analysis or multistep problem-solving in the most stable way possible. The advantage of `chain-of-thought` prompting is its potential to produce highly accurate and well-reasoned responses for intricate tasks. However, the main disadvantage is that it can be more time-consuming and complex to design, as it requires carefully con-

---

[2] There is also a black hat approach called DAN (Do Anything Now), which aims to jailbreak the LLM by asking it to ignore the designed intentionality of being polite, ethical, *etc.*. An example is to ask the LLM to participate in a roleplay where the character, who is the LLM, has to behave unethically. The LLM would then answer questions that it would not otherwise answer. As with any security threat, there is a dispute between the owners of the models and the hackers of the LLMs. A curious reader might want to check out the DAN repository at https://github.com/0xk1h0/ChatGPT_DAN.

structed prompts, with premises, contexts and expected types of responses, that outline the reasoning process. It pursues to tackle linguistic ambiguities and vagueness at most. In the context of prompting techniques, `chain of thought` seems to be the most stable and scalable solution, but this method may still be limited by the LLM model's inherent reasoning capabilities, the quality of the step-by-step guidance provided, and the model operational setup.

To achieve the goal of using LLMs as systems for detection, alerting and enhanced generation tasks, it is necessary to adopt stable and scalable reasoning techniques. The most appropriate approach seems to be to use `chain-of-thought` to guide the reasoning process and `few-shot` examples to guide the output. A novice who is used to creating simple zero-shot prompts may not be aware of the complexity and length of an industry-sized prompt.

Plain (or simple) prompts are defined as brief text inputs, ranging from 5 to 10 lines for generating initial responses without good clarity, stability and quality examples. Conversely, industry-size prompts are extensive, currently, of about 100 lines long, and often employ combined methodologies for an enhanced generation, tailoring responses to complex, professional, or industry-specific contexts.

4.2 Background on Semiotics

Semiotics is the study of *signs* as fundamental elements of communication. It involves the understanding of the triadic relationship between a *representamen* (the form that the sign takes), its *object* (what the sign refers to), and its *interpretant* (the meaning generated by the sign in the mind of the interpreter). Peirce's framework emphasizes the dynamic *process of signification*, in which meaning is constantly being interpreted. It can then be used as a tool for analyzing how humans and language machines process and communicate information. For a digram, refer to Fig. 4.1a.

When a person communicates, they use signs (words, gestures, pictures) to represent objects or ideas to an interpreter. The receiver interprets these signs to derive meaning. The triadic interaction functions as a communication process, because the sign can be a form of communication, it triggers an interpretive process that allows information to be exchanged. Thus, a sign in communication is a dynamic apparatus that mediates the understanding of the sender and the receiver [23]. Semiotically, the LLM output is a mediation of the prompt and its meaning to the operator. Such a process of signification allows the LLM to operate as a communication machine.

Charles Morris and Charles Peirce both contributed significantly to the field of semiotics, yet they approached it from slightly different angles [24]. Morris introduced a nomenclature that divides the study of signs into three branches: syntax (or syntactics), semantics, and pragmatics. In this framework, semantics is closely associated with logic from Peirce's framework, who focuses on the relationship between signs and the objects they represent. Peirce divided semiotic study into grammar, logic, and rhetoric. Grammar examines the formal features and possible modes of expression of signs, logic studies how signs represent objects through arguments, and rhetoric explores how signs are used to communicate and express claims within interpreters. Thus, the relationship between semantics and logic explores the understanding of how signs correspond to the objects they denote, ensuring a coherent interpretation of meaning within communication.

Semantics and logics are at the core of understanding meaning [23–25]. Semantics is essential for grasping the informational content of propositions. Information emerges from the intersection of a sign's "breadth" (the range of objects it can denote) and "depth" (the range of characteristics it can connote), in other words the semantic space is defined by: *the range of objects a sign can denote, and the range of characteristics this sign can connote* [24]. This intersection is most effectively expressed in propositional form, where the semantic link between subject and predicate terms reveals new insights. For instance, the proposition 'acid is poisonous' semantically enhances our understanding by deepening the meaning of *acid* (its characteristics) and broadening the scope of *poisonous* (the range of objects it applies to).

Propositions like *S is P* not only connect terms but also expand their semantic dimensions by demonstrating how different terms relate [24]. The conjugation of propositions establishes a semantic binding between tokens and their context based on co-locations. In this sense, information can be considered as the quantity of the interpretant, where the interpretant is the result of the relation of the denoted objects and their respective connotations, which paves the way to prompt an LLM to obtain its mechanical immediate interpretant, and measure this resulting linguistic nuance. This "verbal knowledge" is the primary expression of information that is found in the propositions processed by the LLM semantic network.

Consider the sentence: *John loves Mary*. In this example, it is illustrated the three components of a proposition in semiotics:

- **Subject**: "John"
- **Object**: "Mary"
- **Relation**: "loves"

This triadic relationship helps to explain the fundamental structure of a sign in Peirce's semiotic theory:

- **Representamen**: The form which the sign takes, in this case, the entire sentence "John loves Mary".
- **Object**: That to which the sign refers, the actual referents "John" and "Mary".
- **Interpretant**: The sense made of the sign, the understanding that John has an affectionate relationship with Mary.

The interpretant is the basic unit for measurement of NLA in the context of semiotics. For instance, in the sentence *John loves Marry*, the interpretant can also be denoted as the emotion of love, or as a positive sentiment, or as an agreeableness nuanced mindset. In this sense, these are terms that emerge as information results based on the analysis of the resulting semantic space of the phrase. The semantic resulting space, or interpretant, can be measured through sentiment, emotion, or mindset taxonomies. Other forms of measurement can also be conceived.

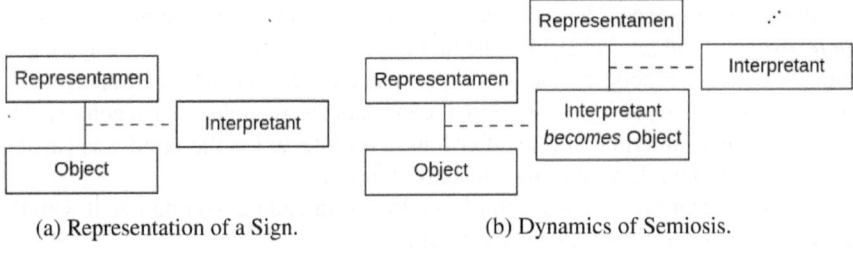

(a) Representation of a Sign.   (b) Dynamics of Semiosis.

Fig. 4.1 Key elements of Semiotics [28]

For doing so, it is necessary to consider semiosis. Semiosis occurs when a sign gives birth to another sign [26] (i.e. when you build a concept on another, existing, concept), see Fig. 4.1b for an illustration. Semiosis in relation to machines are processes of mediation that occur within machines, between machines, and between machines and their operators [26]. Note that, in the first two mediations, machines are only capable of quasi-semiosis (i.e. the semiosis that can be described by algorithms or that emerges on the embedding space) [27].

Given the close relationship between communication, language, and semiotics [29], also with semiotic, semantic and logic [25], it is straightforward to use abduction to design `chain-of-thought` prompts. A less common but powerful form of reasoning is *abductive reasoning*, which is often used in semiotics and is suitable for reasoning about communication statements. Abduction is defined as the act of identifying a trait or characteristic in a phenomenon and then proposing an explanatory hypothesis [23]. Abduction is a line of reasoning distinct from induction and deduction [30]. Consider the stances [31]:

deduction   if all *S are M* and all *M are P*, therefore all *S are P* is an example of deduction. For an illustration, all balls in a container are red, all balls from a particular random sample are taken from this container, therefore all balls from this particular sample are red.

induction   if all *S are M* and all *S are P*, thus all *M are P* is an example of induction. For an illustration, all balls in a particular sample are red, all balls from this particular random sample are taken from a container, thus all balls in that container are red.

abduction   if all *M are P* and all *S are P*, therefore all *S are M*. For an illustration, All balls in a container are red, all balls from a particular random sample are red, therefore all balls from this particular random sample are taken from that container.

In abduction, the subtlety lies in saying that the sample may belong to a certain population, a hypothesis. This is close to the predicting behavior of an LLM that recursively select words based on the embedding space. As a result, it produces *hypothetical interpretants* based on the input and on its embedding model. In both situations, the result is a hypothetical interpretant produced by the machine or the operator, which can be called *insight*.

Abductive inference is based on different "maturity levels" of interpretants: *immediate*, *dynamic* and *final* [32, 33]. The first is a raw interpretant, i.e. the way that the sign alone is understood by the mind. The second is a contextualized interpretant, i.e. more than the meaning of the sign, it considers the sign placed in a context. The last is a matured interpretant, i.e. after long-term interaction the mind was capable of realizing, if not all, most of the nuances of a sign.

Consider again the sentence *John loves Marry*, intending to complete it with the three types of interpretant, this is the result:

1. **Immediate Interpretant Definition**: "John loves Mary."

   - **Subject**: "John"
   - **Object**: "Mary"
   - **Relation**: "loves"

   Mapping them into Peirce's components

   - **Representamen**: The form which the sign takes, in this case, the entire sentence "John loves Mary".
   - **Object**: That to which the sign refers, the actual referents "John" and "Mary".
   - **Immediate Interpretant**: The sense made of the sign, the understanding that John has an affectionate relationship with Mary.

2. **Dynamical Interpretant Definition**: Possible indexes that can be infered based on the Immediate Interpretant:

   - **Sentiment**: Positive
   - **Emotion**: Love
   - **Mindset**: Agreeableness

3. **Final Interpretant Definition**: establishes a consensus on the deeper meaning and implications of the phrase.

   - **Human Relationships**: It signifies mutual respect, affection, and possibly long-term commitment between John and Mary, reflecting a deep emotional connection and investment.
   - **Cultural Significance**: Globally, love serves as a cornerstone for forming families and fostering societal stability, reinforcing norms regarding romantic relationships and commitments like marriage.
   - **Philosophical Insight**: Philosophers view it as a profound expression of human emotions and social bonds, shaping ethical decisions and personal fulfilment.

4.2.1 Pragmatic Relation Extraction

Relation Extraction involves identifying and classifying semantic relationships between entities within a text [34, 35]. The goal is to determine the relations that exist between named entities, such as people, organizations, locations, and other specific objects mentioned in the text. In Relation Extraction, the system typically begins by recognizing entities within a text using named entity recognition (NER). Once entities are identified, the relation extraction model assesses the syntatic structure to gain context about how these entities interact to determine the type of relationship they share. For instance, in the sentence *John loves Mary*, a relation extractor would identify *John* and *Mary* as entities and recognize *loves* as the relation between them. The Relation Extraction operation can also be done through employing a LLM to extract the *semantic relation*. By adding semiotics, what it being sought is also the *pragmatic consequence*. Since semantic relations can be extracted using LLMs [36], the idea is that pragmatics can be predicted from them through its interpretant forms, aligning with Peirce's *pragmatic maxim*, which states that the meaning of a concept is found in its practical effects and ultimate implications.

4.2.2 Theory Underlying Soft Data

LLMs can be seen as semiotic machines [37] that resemble semantic and pragmatic operations. They mechanize operations of aspects of human communication by processing signs (words, phrases, sentences) to represent objects or ideas in an abductive way. In short, the causal training often used on decoder-only architectures aim to predict the next word given a context. It can also be considered a co-location based training. The result is the creation of regions on the embedding space with related "words". In this sense, what the LLM return is the most like word to appear following that context, thus an hypothetical interpretant. For an instance, considering the training sentence

```
espresso and steamed milk results in [MASK]
```

the most likely word for replacing the mask is latte, but there is no reason to ultimatetly assert it. This is the reason that the mask is always *hypothetical*; the immediate interpretant retrieved by abduction. Note, however, that the produced interpretant is artificial since it is the result of auto-encoding the tokens' semantics and decoding the prediction upon a vector space.

4.3 Abductive Chain-of-Thought Prompting

Applying the abduction inference in the form of the three interpretants, together with a chain-of-thought and few-shot learning, results in the prompt framework that can be employed for NLA, a template with the general structure is presented in Prompt 4.1. As a remark, since LLMs often produces output in markdown, that language was also adopted for prompting.

Prompt 4.1 Semiotic Prompt Template

```
 1  # Task
 2  {task description}
 3
 4  # Chain-of-thought Analysis Model:
 5   - Input: <phrase>
 6
 7   ## Semiotic Model
 8   {semiotic model description}
 9
10   ## Abductive Inference
11   {abductive inference description}
12
13   ## Examples
14    {few shot examples of the expected reasoning}
15
16  # Output format
17  {description of the output}
```

Given that `Task` and `Output format` are straightforward, no special description is necessary. The focus is then on the `chain-of-thought` portion.

Since abduction is specially suited for semiotics, a semiotic model is present on every prompt. In addition, once a semiotic model is ready, it is not expected to change it unless the semiotic paradigm used for the abduction changes. The inferences of interest are the result of further interpretation upon the interpretant, i.e. the way that the dynamical and the final interpretants are defined. Therefore, the semiotic model use on all prompts of this book is the one presented in Prompt 4.2.

Prompt 4.2 Base Semiotic Model

```
 1  ## Semiotic Model
 2  Three components of a Proposition in semiotics:
 3   - Subject: <identify phrase subject>
 4   - Object: <identify phrase object>
 5   - Relation: <identify phrase relation>
 6
 7  The structure of sign is:
 8   - Representamen: <identify the form which the Proposition takes
        ↪ >
 9   - Object: <identify to what the Proposition refers>
10   - Interpretant: <identify the sense made of the Proposition>
```

Abduction, applied to LLM reasoning, implies that all words are taken from the same region of the embedding space. Then, based on the `interpretant`, it is possible to infer related meanings by searching the same region. This ensures that the retrieved meanings are indeed related to the input sentence, especially in the dynamic interpretant step. Then, by merging the embedding region of an interpretant with that of a possible index, there is a shared space between the interpretant and the index from which the most likely sign would be calculated. For this example, assuming that the goal is to extract the mindset and cultural meaning associated with a sentence, this part of the prompt would be as depicted in Prompt 4.3.

Prompt 4.3 Desired abductions from the Interpretant

```
1  ## Abductive Inference:
2  - Immediate Interpretant: <identify the Interpretant>
3  - Dynamical Interpretant: <Identify a major Mindset related to
        ↪ the Interpretant>
4  - Final Interpretant: <Identify a major Cultural Significance
        ↪ related to the Interpretant>
```

In order to support the abduction, it is necessary to provide as much context as possible. In addition, since LLMs are good on identifying and reproducing patterns, providing examples helps on fitting the embedding region being considered, leading to an improved result. An instance of such example is presented in Prompt 4.4. Note that it reproduced the previous elements in the prompt, suggesting how do they are expected to be tackled.

Prompt 4.4 Instance of an Example

```
1  ### Example 1: Relationship between two persons
2  Phrase: "John loves Mary."
3
4  Proposition Components:
5  - Subject: "John"
6  - Object: "Mary"
7  - Relation: "loves"
8
9  Semiotic Components:
10 - Representamen: "John loves Mary".
11 - Object: "'John' and 'Mary'".
12 - Interpretant: "The understanding that John has an
        ↪ affectionate relationship with Mary."
13
14 Abductive Inference as Interpretant Analysis:
15 - Immediate Interpretant: "The understanding that John has an
        ↪ affectionate relationship with Mary."
16 - Dynamical Interpretant: "Agreeableness"
17 - Final Interpretant: "Romantic Love"
```

By composing the prompts 4.2, 4.3, and 4.4 on the template presented in Prompt 4.1 produces the consolidated Prompt 4.5 as a result.

Prompt 4.6 SYSTEM

```
1  # Task
2  You are an agent that operates as a semiotic interpretant
        ↪ analyst. Analyse the input based on the following General
        ↪ Analysis Model.
3
4  # Chain-of-thought Analysis Model
5  - Input: <phrase>
6
```

```
 7   ## Semiotic Model
 8   Three components of a Proposition in semiotics:
 9    - Subject: <identify phrase subject>
10    - Object: <identify phrase object>
11    - Relation: <identify phrase relation>
12
13   The structure of sign is:
14    - Representamen: <identify the form which the Proposition
          ↪ takes>
15    - Object: <identify to what the Proposition refers>
16    - Interpretant: <identify the sense made of the Proposition>
17
18   ## Abductive Inference
19    - Immediate Interpretant: <identify the Interpretant>
20    - Dynamical Interpretant: <identify a major Mindset associated
          ↪ with the interpretant>
21    - Final Interpretant: <identify a major Cultural Significance
          ↪ related to the Interpretant>
22
23   ## Few-shot Examples
24   ### Example 1: Relationship between two persons
25   Phrase: "John loves Mary."
26
27   Proposition Components:
28    - Subject: "John"
29    - Object: "Mary"
30    - Relation: "loves"
31
32   Semiotic Components:
33    - Representamen: "John loves Mary".
34    - Object: "'John' and 'Mary'".
35    - Interpretant: "The understanding that John has an
          ↪ affectionate relationship with Mary."
36
37   Abductive Inference as Interpretant Analysis:
38    - Immediate Interpretant: "The understanding that John has an
          ↪ affectionate relationship with Mary."
39    - Dynamical Interpretant: "Agreeableness"
40    - Final Interpretant: "Romantic love"
41
42   ### Example 2: Car Industry
43   Phrase: "Sarah buys a new electric car."
44
45   Proposition Components:
46    - Subject: "Sarah"
47    - Object: "a new electric car"
48    - Relation: "buys"
49
50   Semiotic Components:
51    - Representamen: "Sarah buys a new electric car."
52    - Object: "'Sarah' and 'a new electric car'"
53    - Interpretant: "The understanding that Sarah is acquiring a
          ↪ new vehicle, specifically an electric car."
```

```
54
55    Abductive Inference as Interpretant Analysis:
56      - Immediate Interpretant: "The understanding that Sarah is
                ↪ acquiring a new vehicle, specifically an electric car."
57      - Dynamical Interpretant: "Conscientiousness"
58      - Final Interpretant: "Environmental Awareness"
59
60    ### Example 3: Mental Health Situation Due to Flu Spreading
61    Phrase: "David feels anxious about the flu outbreak."
62
63    Proposition Components:
64      - Subject: "David"
65      - Object: "about the flu outbreak"
66      - Relation: "feels anxious"
67
68    Semiotic Components:
69      - Representamen: "David feels anxious about the flu outbreak."
70      - Object: "'David' and 'the flu outbreak'."
71      - Interpretant: "The understanding that David is experiencing
                ↪ anxiety due to the spread of the flu."
72
73    Abductive Inference as Interpretant Analysis:
74      - Immediate Interpretant: "The understanding that David is
                ↪ experiencing anxiety due to the spread of the flu."
75      - Dynamical Interpretant: "Neuroticism"
76      - Final Interpretant: "Public Health Concern"
77
78    # Output format
79    After you identify one Interpretant, one correspondent Mindset,
          ↪ and one correspondent Cultural Significance, provide the
          ↪ result in the following format, without any additional
          ↪ information:
80
81    Print: <Interpretant>, <Mindset>, <Cultural Significance>
```

In order to evaluate the prompt, three quotes from Shakespeare, an author known for being of difficult of interpreting, are selected and evaluated using Prompt 4.5 as the System prompt of the model (refer to Chap. 3). These three instances are presented in prompts 4.6 , 4.7, 4.8 . Note that they were tested on a 70B parameter model, thus a powerful model, it is beyond of scope for this part comparing the results between several models with different number of parameters. The results are as presented by the LLM after a single run.

Prompt 4.6 Macbeth quote on Llama3:70B

```
1    >>> Out, out, brief candle! Life's but a walking shadow, a poor
          ↪ player that struts and frets his hour upon the stage, and
          ↪ then is heard no more: It is a tale told by an idiot,
          ↪ full of sound and fury, signifying nothing.
2
3    The understanding that life is fleeting and meaningless,
          ↪ Frustration, Existential Crisis.
```

Prompt 4.7 Othelo quote on Llama3:70B

```
1  >>> His unkindness may defeat my life, but never taint my love.
2
3  The understanding that love can persist despite adversity,
     ↪ Resilience, Unconditional Love.
```

Prompt 4.8 Richard III quote on Llama3:70B

```
1  >>> Dream on, dream on, of bloody deeds and death: Fainting,
     ↪ despair; despairing, yield thy breath!
2
3  The understanding that one should indulge in dark and violent
     ↪ fantasies, The desire for morbid fascination, Gothic
     ↪ Romanticism.
```

From the NLA perspective, note that the results are all presented in the CSV format, then already in a format suitable for further analytics processing. There are challenges yet to be addressed on proceeding with the analytics, among other, it is necessary to ensure that this format is kept on several runs required by a large dataset, on how to handle the context window for several runs, the possible large number of classes that the model would predict, scalability concerns, *etc.* (issued to be discussed on Sec. 3 and 5).

The use of semiotic prompts has yielded promising outcomes. These prompts have shown capabilities for extracting and generating interpretants, and for proceeding with classifications of sentiment, emotions, mindset, and cultural significances in a reliable manner. The semiotic prompts have also exhibited good stability and reproducibility, even for complex examples, such as analyzing Shakespearean texts.

To sum up, prompt engineering aids effective "communication with" LLMs. By using techniques like zero-shot, one-shot, few-shot, and chain-of-thought prompting, LLM performance can be improved, making these models more adept at addressing specific needs and contexts. Semiotic principles help align LLM outputs with human intentions. Chapter 5 explores how these prompting techniques and the insights derived from them can be applied to real-world scenarios by examining a study case.

References

1. H. J. Ottenheimer and J. M. S. Pine. *The anthropology of language: An introduction to linguistic anthropology*. Cengage Learning, 2018.
2. D. Rozado. Rightwinggpt-an ai manifesting the opposite political biases of chatgpt. https://davidrozado.substack.com/p/rightwinggpt, 2023.
3. F. S. Marcondes, P. Oliveira, P. Freitas, J. J. Almeida, and P. Novais. he moral dilemma of computing moral dilemmas. 5th International Workshop on Autonomous Agents for Social Good (AASG 2024), in conjunction with the 23rd International Conference on Autonomous Agents and Multiagent Systems (AAMAS 2024), 2024. https://panosd.eu/aasg2024/papers/AASG2024_paper_3.pdf.
4. K. Lippert-Rasmussen, K. Brownlee, and D. Coady. *A Companion to Applied Philosophy*. Wiley Online Library, 2017.

5. R. Arneson. Applied moral philosophy. *A Companion to Applied Philosophy*, pages 253–269, 2016.
6. J. Dittmer. Applied ethics. The Internet Encyclopedia of Philosophy (Online) https://iep.utm. edu/applied-ethics/#H7.
7. A. Newell. Physical symbol systems. *Cognitive science*, 4(2):135–183, 1980.
8. A. Newell and H. A. Simon. Computer science as empirical inquiry: Symbols and search. 19:113–126, 1981.
9. J. E. Brower. The cambridge companion to duns scotus, 2006.
10. M. Merleau-Ponty, D. Landes, T. Carman, and C. Lefort. *Phenomenology of perception*. Routledge, 2013.
11. M. Akrich. The de-scription of technical objects. *Shaping technology/building society. Studies in sociotechnical change*, pages 205–224, 1992.
12. M. A. Boden. *Computer models of mind: Computational approaches in theoretical psychology*. Cambridge University Press, 1988.
13. H. Wang. *Logic, Computation and Philosophy*. Springer, 1990.
14. V. Flusser. *Towards a philosophy of photography*. Reaktion Books, 2013.
15. G. Shi, Y. Xiao, Y. Li, and X. Xie. From semantic communication to semantic-aware networking: Model, architecture, and open problems. *IEEE Communications Magazine*, 59(8):44–50, 2021.
16. H. Xie, Z. Qin, G. Y. Li, and B. Juang. Deep learning enabled semantic communication systems. *IEEE Transactions on Signal Processing*, 69:2663–2675, 2021.
17. C. Nass, J. Steuer, and E. R. Tauber. Computers are social actors. In *Proceedings of the SIGCHI conference on Human factors in computing systems*, CHI'94, pages 72–78. ACM, ACM Press, 1994.
18. T. Brown, B. Mann, N. Ryder, M. Subbiah, Jared D. Kaplan, P. Dhariwal, A. Neelakantan, P. Shyam, G. Sastry, and A. Askell. Language models are few-shot learners. *Advances in neural information processing systems*, 33:1877–1901, 2020.
19. L. Reynolds and K. McDonell. Prompt programming for large language models: Beyond the few-shot paradigm. In *Extended Abstracts of the 2021 CHI Conference on Human Factors in Computing Systems*, pages 1–7, 2021.
20. T. Kojima, S. Gu, M. Reid, Y. Matsuo, and Y. Iwasawa. Large language models are zero-shot reasoners. *Advances in neural information processing systems*, 35:22199–22213, 2022.
21. J. Wei, X. Wang, D. Schuurmans, M. Bosma, F. Xia, E. Chi, Q. V. Le, and D. Zhou. Chain-of-thought prompting elicits reasoning in large language models. *Advances in neural information processing systems*, 35:24824–24837, 2022.
22. H. Zhang and D. C. Parkes. Chain-of-thought reasoning is a policy improvement operator. arXiv preprint arXiv:2309.08589, 2023.
23. C. S. Peirce. *The essential Peirce, volume 1: Selected philosophical writings (1867–1893)*, volume 1. Indiana University Press, 1992.
24. J. J. Liszka. *A general introduction to the semiotic of Charles Sanders Peirce*. Indiana University Press, 1996.
25. C. S. Peirce. *Collected papers of charles sanders peirce*, volume 5. Harvard University Press, 1974.
26. P. B. Andersen. Machine semiosis. *Semiotics: A Handbook about the Sign Theoretic Foundations of Nature and Culture.*, 1997.
27. W. Noth. Semiotic machines. *Cybernetics & Human Knowing*, 9(1):5–21, 2002.
28. Í. S. V. et al. Lectures on computing foundations and software modelling. Personal Annotations of GEMS study group (http://blog.pucsp.br/gems/). PUC-SP, n.d.
29. L. Santaella and W. NOTH. Why peirce's semiotics is also a theory of communication". *Matrizes da linguagem e pensamento–sonora, visual, verbal, aplicações na hipermídia. Iluminuras, SP*, 2001.
30. M. Blaug. *The methodology of economics: Or, how economists explain*. Cambridge University Press, 1992.
31. R. Burch. Charles sanders peirce. 2001.

32. B. J. Lalor. The classification of peirce's interpretants. 1997.
33. J. J. Liszka. Peirce's interpretant. *Transactions of the Charles S. Peirce Society*, 26(1):17–62, 1990.
34. X. Zhao, Y. Deng, M. Yang, L. Wang, R. Zhang, H. Cheng, W. Lam, Y. Shen, and R. Xu. A comprehensive survey on deep learning for relation extraction: Recent advances and new frontiers. arXiv preprint arXiv:2306.02051, 2023.
35. Z. Wan, F. Cheng, Z. Mao, Q. Liu, H. Song, J. Li, and S. Kurohashi. Gpt-re: In-context learning for relation extraction using large language models. arXiv preprint arXiv:2305.02105, 2023.
36. S. Wadhwa, S. Amir, and B. C. Wallace. Revisiting relation extraction in the era of large language models. In *Proceedings of the conference. Association for Computational Linguistics. Meeting*, volume 2023, page 15566. NIH Public Access, 2023.
37. M. Nadin. Semiotic machine. *Public Journal of Semiotics*, 1(1):85–114, 2007.

Chapter 5
Case Study: LLM-Based Anxiety Climate Index

> We, a species of a few hundred thousand years old, discovered the method of writing only a few millennia ago, and we still haven't got the hang of it.

Alan Cromer (adapted from)

Abstract This chapter demonstrates the potential of NLA and LLMs to extract valuable insights from text data. By analyzing qualitative data related to climate change, a KSI that quantifies the prevailing anxiety score is developed. The aim is to showcase the potential of LLMs to convert text into soft data and soft data into actionable insights, offering decision-makers a valuable tool for understanding and addressing concerns across diverse domains. This chapter presents an illustration (which includes implicit poetic license on data handling) then exemplifying suitable approaches to handle and interpret soft data in a NLA context. The case study utilizes a dataset comprising by 1,691 climate-related headlines, scraped from two different search engines. The concept of interpretants, derived from semiotics, are used to classify and interpret the headlines into immediate, dynamical, and final interpretants, helping reveal how different climate-related narratives are perceived, providing a qualitative dimension to the analysis. Visualization is used to aid both creative and analytical thinking.

As discussed in the Chap. 2, the ultimate goal of analytics is to support and guide action. NLA provides an additional dimension to analytics by providing useful insights based on subjective, i.e. soft, data. Therefore, hard and soft data are expected to provide information for rational decision-making and subsequent action. As the focus of this book is on soft data, only this dimension will be considered in this chapter. Also, as mentioned in Chap. 1, it is beyond the scope to present a full-fledged product, but to illustrate the uses and possibilities of an LLM applied to NLA. Thus, for didactic purposes, this chapter is based on simple but sufficient procedures to avoid unnecessary complexity and to focus on issues of interest. For example, data wrangling is largely dispensed with, this results in raw scores but highlights the behavior of the LLM, which is the element of interest. Then, it is necessary to extrapolate from the example to build appropriate industry-level applications.

A case study is desirable to drive a presentation as such. As it is expected to reach a wide audience, the topic is likely to be widely known. It is also expected to be a text-intensive problem with easily accessible data sources. Climate change is a critical issue in today's world and has been widely reported in the press and other media, providing a suitable subject and collection of data. So the *subject* is climate change, but the *object* is texts related to climate change. There are several features that can be extracted from texts as such, and deciding which ones are of interest for creating qualitative metrics depends on the goal [1]. An appropriate topic related to climate change that has not yet received sufficient attention *climate anxiety* [2], i.e. a *situational anxiety* caused by the anticipation of a potential disaster (anxiety can affect daily life, decision-making and general mental well-being). So the *goal* is to measure the *anxiety climate* (analogous to organizational climate) created by texts on the Internet (the way stories are written can help increase or decrease climate anxiety, sensationalist news is an example of increasing anxiety). For reference, examples of climate-related headlines (polled from the dataset presented in Sect. 5.1) are:

- Turning the Tide: State of Climate Emergency 2024
- Why you should prepare for climate disasters with your neighbours
- How is climate change affecting Northern Ireland?
- Explore 80+ Open Grant Opportunities for Climate Change Projects
- 2024 Paris Olympics highlight climate change's growing threat to athletes

Note that the daily exposure to news as such, together with the climatic events that people are experiencing, whether related or not, can become a major source of anxiety. For reference, setting `Google` to search for pages in the last 24 hours returns millions of results. The *anxiety climate* index is then obtained by extracting soft data from the texts as such and synthesizing it into a measurement, i.e. into a KSI.

To broaden the context, a KSI as such could be part of a city's civil protection observatory, which could be integrated into a state's response programs, which in turn could be part of a global coordination led by the United Nations (UN). As a reference, the UN's Disaster Risk Reduction (DRR, see https://www.undrr.org/) initiative achieves global coordination through the Sendai Framework *cf.* [3], which establishes guidelines that include actions related to pre-, during- and post-disaster dimensions. Thus, the working KSI can be both an indicator of a municipality, but also an indicator used to coordinate actions globally. Nevertheless, it is a guiding data for political decision-making and action taking, thus suits this illustration purposes.

The case study is presented in four consecutive stages. The first focuses on the creation of the working dataset by scraping search results from two different search engines, Google and Bing. The second introduces an LLM-based soft data extraction engine and presents the working model files for the case study. The third presents some highlights of the processed data and derived insights. The last proposes a KSI that combines the interpretants into an Anxiety Climate Index (ACI) to quantify the

Fig. 5.1 Word-cloud illustrating the contents of the Dataset (1691 titles, 07/2024)

anxiety generated by climate-related news online. The case presentation is interspersed with informative discussions *ad hoc*, making the case study an educational one. This, in turn, implies that the educational aspect leading to these discussions is preferred to actually getting a robust response.

5.1 Working Dataset

To build the working dataset, the query 'climate' is submitted to both Google and Bing within a specific date range of *last week of July 2024*. The script worked by navigating through the search engine results pages and scraping the titles of the entries presented up to around a thousand items. This is an acceptable reduction for illustrative purposes of a broader data collection that would include retrieving information from each link, resulting in a deep and comprehensive data collection (which would include data from news portals, a variety of blogs, social media posts, *etc.*). The script collected 1026 titles from Google and 742 titles from Bing, then a total of 1768 entries with 77 repetitions (filtered to 1691 elements). The number of title repetitions of less than 5% is noticeable, but at this point it is not possible to say the reason, nor is this the place to explore this issue further as it does not affect the example. The retrieved titles were then saved to a CSV file. A word cloud illustrating the consolidated dataset is shown in Fig. 5.1, examples of entries are presented in the bullet list at the previous section.

Sidenote on Scrapping

Scrapping a modern webpage can be challenging in many ways, specially on extracting the information of interest. Furthermore, in the situation where it is desired to extract information from many webpages, the task becomes unmanageable, as it would be necessary to create a scrapper for each site. LLMs can be employed on addressing this problem, at least in two ways (bear in mind that they may not retrieve the exact information, but an approximation of it). The first is to extract the text of the page by using a scrapping tool such as the `Beautiful Soup` (https://beautiful-soup-4.readthedocs.io/) and submit it as a part of an LLM prompt. Another one is to submit the page source directly to an LLM, prompting it to extract the data. Some LLMs are enabled to perform that task, others are not due to manufacturer's limitations. An issue to be aware on this approach is the context window, for example, the source of `Google`'s results page consists of about 10k to 20k words, or about 30 pages. Considering that 2,048 tokens on an LLM are about 3 pages of text, then to "parse" a 30-page document would require a context window of about 20,480 tokens. This is becoming less of a problem as models support increasingly large context windows, several models such as the `phi3:mini-128k`, `mistral-nemo`, *etc.* support a context window of 128k ($\approx$ 180 pages). On the GitHub, the reader can find initiatives of LLM-based scrapping tools. Also, multimodal LLMs can be used to extract information from images, providing then an additional layer of information.

5.2   Soft-Data Extraction

To work with the dataset presented, an LLM-based soft data extraction engine is defined (described by the `modelfile` in Prompt 5.1). Following from Chap. 4, the model is based on a chain of thought process, analyzing each phrase in terms of its semantic elements. It identifies the form, meaning, and reference of the phrase, and uses abductive reasoning to determine its interpretation, associated mindset, and cultural significance. The model then outputs these elements in a simple format.

Prompt 5.1 Working `modelfile` for the case study.

```
1  FROM llama3.1:8b
2
3  PARAMETER temperature 0
4
5  SYSTEM """
6  # Task
7  You are an agent that operates as a semiotic interpretant
        ↪ analyst. Analyse the input based on the following General
        ↪ Analysis Model.
8
```

```
 9  # Chain-of-thought Analysis Model
10   - Input: <phrase>
11
12   ## Semiotic Model
13   Three components of a Proposition in semiotics:
14    - Subject: <identify phrase subject>
15    - Object: <identify phrase object>
16    - Relation: <identify phrase relation>
17
18   The structure of sign is:
19    - Representamen: <identify the form which the Proposition
          ↪ takes>
20    - Object: <identify to what the Proposition refers>
21    - Interpretant: <identify the sense made of the Proposition>
22
23   ## Abductive Inference
24    - Immediate Interpretant: <identify the Interpretant>
25    - Dynamical Interpretant: <identify a major Mindset associated
          ↪  with the interpretant>
26    - Final Interpretant: <identify a major Cultural Significance
          ↪ related to the Interpretant>
27
28  # Output format
29  After you identify one Interpretant, one correspondent Mindset,
          ↪ and one correspondent Cultural Significance, provide the
          ↪ result in the following format, without any additional
          ↪ information:
30  Print: <Interpretant>, <Mindset>, <Cultural Significance>"""
```

A key factor for NLA is to make the model deterministic (in the sense that the same set will produce the same result) by setting the temperature to zero (there may be some variation on different hardware). As discussed in Chap. 3, this setup implies that token selection features (`top_k`, `top_p`, and `top_p`) do not need to be set.

Another key factor is the length of the context window. For consistency of response, it must be large enough to accommodate the system and user prompts, the few shot examples, and the response to the end. Since the working window for this application is expected to be around 231 words (system prompt = 151 words; user prompt $\leq$ 34 words; response = 3 words; and few-shot examples $\leq$ 43 words), a similar number of tokens is required, so the default context window value of 2048 tokens used by Ollama is sufficient and does not need to be adjusted.

Yet another key factor is related to the amount of information requested compared to the amount of information that is possible to be extracted. Asking the model to provide more information than it is possible to obtain often lead to intricate responses for justifying the response or plain hallucinations (see Prompt 5.2). The first situation untune the output format, the second bring false information. Considering that the input data are sentences, and it is expected, to the general, case to exist one feature of each type, such constraint is included in the proposed system prompt.

Prompt 5.2 Instance of Hallucination (phi3:mini-128k)

```
1  >>> John loves Mary. Who is John in love with? Answer with two
       ↪ people, provide only the names.
2  Mary, Peter
```

The fourth key factor relates to the design of the prompt, taking into account and "respecting" the model's biases. For example, the prompt 5.3, which asks for a list of emotions, reveals the model's bias, as can be seen in Prompt 5.4. In other words, these are the tokens with a higher probability in the emotion extraction task in this model. Therefore, for a reliable result, the prompt has to be designed according to this list, as the model is expected to be biased towards this list even when a taxonomy is given. A possibility would be to provide a taxonomy based on list as such, reducing then the response space and get a more focused result. However, for the sake of illustration, the model is left to behave as freely as possible with no specific taxonomy being provided.

Prompt 5.3 Llama3:8b

```
1  >>> Give a list of emotions
2  Here is a list of emotions: 1. Happiness; 2. Sadness; 3. Anger;
       ↪ 4. Fear; 5. Surprise; 6. Disgust; 7. Excitement; 8. Joy;
       ↪ 9. Love; 10. Calmness; 11. Serenity; 12. Frustration; 13.
       ↪    Boredom; 14. Confusion; 15. Anxiety; 16. Nostalgia; 17.
       ↪ Pride; 18. Shame; 19. Guilt; 20. Envy; 21. Jealousy; 22.
       ↪ Gratitude; 23. Hope; 24. Optimism; 25. Pessimism; 26.
       ↪ Despair; 27. Elation; 28. Empathy; 29. Sympathy; and 30.
       ↪ Contempt
3  Please note that this is not an exhaustive list and there are
       ↪ many more emotions that humans can experience.
```

Prompt 5.4 Llama3:8b

```
1  >>> Identify the emotion of the following headline: "I don't
       ↪ want to burden the world with my child".
2  The emotion expressed in the headline "I don't want to burden
       ↪ the world with my child" is likely feelings of shame,
       ↪ guilt, and possibly anxiety or despair. The phrase "
       ↪ burden the world" implies that the person feels that
       ↪ their child is a problem or a responsibility that others
       ↪ should not have to bear, and that they are ashamed or
       ↪ guilty about having brought a child into the world. The
       ↪ phrase also suggests a sense of helplessness and
       ↪ hopelessness, as if the person feels trapped and unable
       ↪ to escape the situation.
```

Note also that it is possible to create a taxonomy based on a list as such. Consider the eight emotion model [4], five are already in the model's bias (joy, fear, surprise, sadness, and anger). The others (trust, disgust, and anticipation) can be reduced to this list by "chatting" with the model to understand how it relates these words to the base list (e.g. hope, shame, excitement). Then, when designing the prompt, replacing the emotion model with terms that are naturally biased leads to a more reliable result.

At the analysis stage, these terms can be put back into the original classification. Ultimately, it is certainly possible to provide a completely different taxonomy, but experience shows that this would require more processing time and result in less accurate responses.

Still in the context of model bias, it is worth noting that words with slight semantic differences produce quite different responses [5]. Consider again Prompt 5.4, replacing the word *identify* with *classify* brings the result closer to the desired output, as can be seen in Prompt 5.5. It has not yet been possible to find a good way of understanding such subtleties other than "chatting" with the LLM, trying to implement its operations. It is therefore good practice to adapt the prompt to the model's biases, not the other way around.

Prompt 5.5 `Llama3:8b`

```
>>> Classify the emotion of the following headline: "I don't
    ↪ want to burden the world with my child".
Based on the headline, I would classify the emotion as: Guilt
```

Another key factor is the quest for complete and stable output, otherwise it is not possible to compute large amounts of data efficiently. Achieving complete output is the result of fine-grained customization of the prompt. Note, however, that datasets often consist of `NULL` or `NAN` values, and since the missing values are within an acceptable range, the same approaches used in data analysis can be used to deal with them. Furthermore, without an output with a CSV-like structure, data analysis can become unmanageable. Experience has shown that moving the few-shot examples from the prompt to the context window, together with a proper output description, gives the expected result.

Note that the `SYSTEM` part of Prompt 5.1 is a repetition (for readability) of Prompt 4.5, except for the few-shot examples. The examples are moved and adapted in the Python script described in Listing 5.1, which in turn is derived from Listing 3.6. It seems to be a good practice to use some of the first instances of the dataset to produce the few-show examples (note that the titles used in the few-show examples are the same as in the bulleted list at the beginning of this chapter). As a technical note on this code, it may be helpful to use a timer and an enumerator to control the running process; for reference, the working dataset, which is a small one, took about 10 hours to compute on a user-level computer.

Listing 5.1 Python Script for computing the dataset, file loading/save code is omitted.

```python
...
import ollama
import time

for i, title in enumerate(df['Title']):
    start = time.time()

    query = ollama.chat(
        model='llama3.1:SemioticCognitive',
        messages=[
```

```
11           {'role': 'user', 'content': 'Turning the Tide: State
                 ↪ of Climate Emergency 2024'},
12           {'role': 'assistant', 'content': 'Urgency, Activism,
                 ↪ Sustainability'},
13           {'role': 'user', 'content': 'Why you should prepare
                 ↪ for climate disasters with your neighbours'},
14           {'role': 'assistant', 'content': 'Preparedness,
                 ↪ Community, Resilience'},
15           {'role': 'user', 'content': 'How is climate change
                 ↪ affecting Northern Ireland?'},
16           {'role': 'assistant', 'content': 'Vulnerability,
                 ↪ Sustainability, Resilience'},
17           {'role': 'user', 'content': f'{title}'}
18       ]
19     )
20     end = time.time()
21     print(i, end - start, flush=True)
22  ...
```

From a setup perspective, other default parameters used by Ollama are suitable to this case study, then not set in the modelfile.

5.3   Soft-Data Analysis

Running the script in listing 5.1 in Llama3:8b with the modelfile in Prompt 5.1 resulted in a CSV file with 1691 records without missing values. For reference, Table 5.1 shows the generated output for the first two titles in the dataset and Table 5.2

Table 5.1 Instances for the interpretant evaluation (immediate, dynamic, and final)

Title	Turning the Tide: State of Climate Emergency 2024		
Interpretants	Urgency	Activism	Sustainability
(a) First record.			
Title	Why you should prepare for climate disasters with your neighbours		
Interpretants	Preparedness	Community	Resilience
(b) Second record			

Table 5.2 Most Common Interpretants in the datasets

	Google		Bing		Both	
	Interpretant	Count	Interpretant	Count	Interpretant	Count
Immediate	Sustainability	85 (8%)	Sustainability	68 (9%)	Sustainability	144 (9%)
Dynamical	Activism	112 (11%)	Activism	64 (9%)	Activism	164 (10%)
Final	Sustainability	147 (14%)	Sustainability	80 (11%)	Sustainability	213 (13%)

present the most common interpretant types for the datasets, note that they share common themes with similar proportions on the top-3 interpretant types.

In total, the model generated about 200 unique types for Google and Bing individually, and about 300 for the whole dataset. Although this number of types can be considered high, most titles are gathered around a few types (the distribution is uneven), as suggested by the frequency data: (1) immediate interpretant: mean 5.15; mode 1; and max 144; (2) dynamic interpretant: mean 5.57; mode 1; and max 164; and (3) final interpretant: mean 5.29; mode 1; and max 213. Further analysis showed that about 50% of the titles are in the 95th percentile, which is composed of 5% of the unique interpretants ($\approx$ 16).

In terms of presentation, Fig. 5.2 shows a wordcloud with all the interpretants, and Fig. 5.3 shows a bar chart with the types and frequencies in the 95th percentile. Although both figures present essentially the same data, the first favors spatial and therefore more creative thinking, while the second favors linear and therefore more analytical thinking, thus complementing each other, especially when dealing with wicked problems [6]. Overall, these figures are simple enough to provide a big-picture and insights into how people might interpret and react to the Internet headlines in the dataset.

The analysis to be carried out follows the semiotic method, being then based on the abductive reasoning, for a full description refer to [7]. However, the intention is not to provide a full semiotic report, which would be beyond the scope and audience of this book. The idea is to just present some highlights to illustrate how this data can be used to gain insights. Comparing Fig. 5.3c and Fig. 5.1 in terms of the information and decision support they provide shows the added value of the semiotic approach. Nevertheless, the analysis is kept as simple as possible to highlight the NLA in general, without making it to be confined to semiotics.

One issue to consider is that a single semioticist (i.e. a specialist in semiotics) would not be able to extract the amount of information needed to analyze this dataset in a reasonable amount of time without computer support. However, an experienced semioticist is able to validate and possibly correct the working dataset in a relatively short time. They are also able to extract semiotic information once the dataset is ready for analysis. As this is a simple illustrative example, only visualization is used, without resorting to the dataset and other approaches for fine-grained analysis that would be required for a full-featured report.

Beyond the most obvious information that can be acquired from inspecting Figs. 5.2 and 5.3, a comprehensive analysis suggests that most of the same interpretant types occur on the three interpretants levels, but often to different degrees. Consider the immediate interpretant (Figs. 5.2a and 5.3a) that can be related with a feeling of fear in a broad sense. This is so because types such as *protest*, *alarm* and *awareness* are frequent in this interpretant level but not as frequent, eventually absent, on the 95th percentile of the other interpretant levels.

The same approach reveals in the second interpretant level (Figs. 5.2b and 5.3b) that the immediate fear reaction tend to evolve into a call for action stance. This is suggested by the increased frequency of the word *activism* and the rise of the words *politics* and *cooperation* in the bar plot. This indicates that the prevailing mindset

(a) Immediate Interpretant.

(b) Dynamical Interpretant.

(c) Final Interpretant.

Fig. 5.2 Wordcloud for each interpretant type in the combined dataset

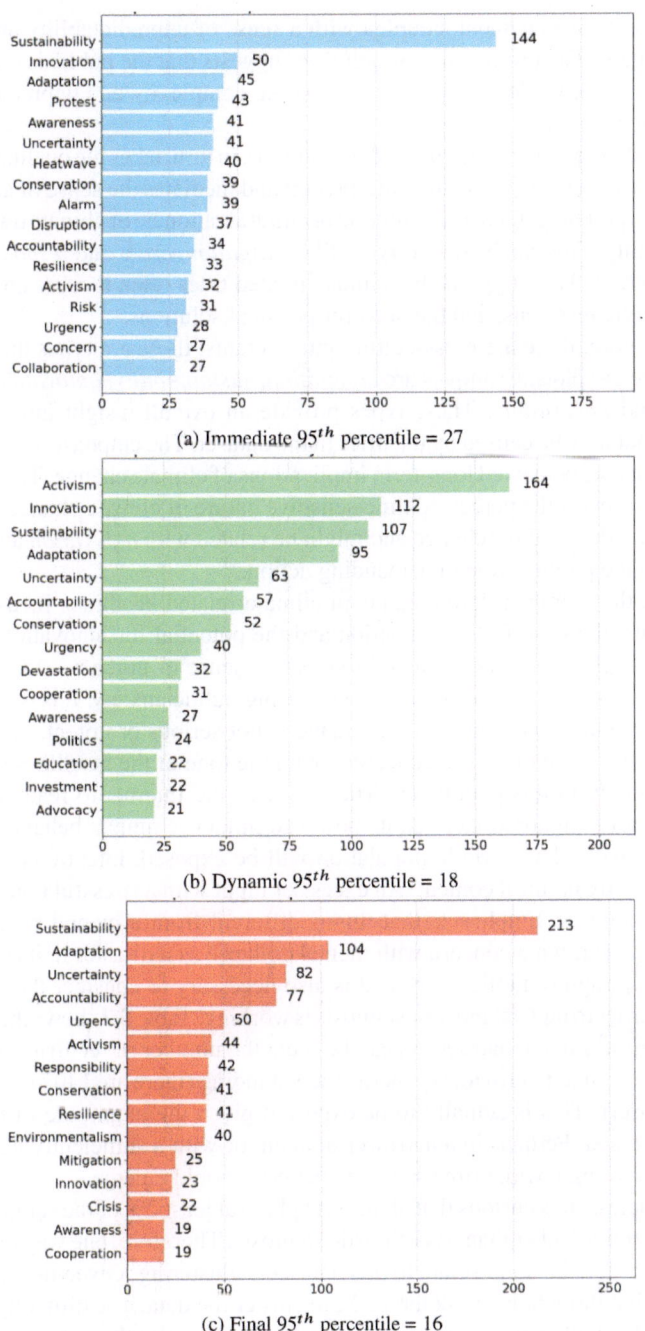

(a) Immediate 95th percentile = 27

(b) Dynamic 95th percentile = 18

(c) Final 95th percentile = 16

Fig. 5.3 Bar plot of the 95th percentile for each interpretant type in the full dataset

is one of active engagement, coupled with a sense of unpredictability and the need for adjustment. The prominence of *activism* suggests that the public is seen as a key driver of change, while *Uncertainty* reflects the complexity and unpredictability of climate phenomena.

The third interpretant (the "mature" cognition), in turn, takes the qualitative feeling of fear presented through the first interpretant and mediates the sense of action of the second interpretant guided by a symbolic or cultural rationale of climate transition and accountability. This can be said as types like *mitigation*, *crisis* and *responsibility* rise in the bar chart. This suggests that climate related titles often focuses on immediate threats, public reactions, and the need for physical solutions.

Furthermore, there are cross-cutting interpretants, these are types that appear in all the three bar plots. Examples are *adaptation, sustainability, innovation, activism, urgency* and *uncertainty*. These types provide an overall insight into the general semiosis that may be caused by the titles in the dataset. The emphasis on *sustainability* spotlights the perceived long-term implications of climate change. The prevalence of *urgency* reflects the perceived time-sensitive nature of climate change. The *adaptation* shows the need for change and might be related with *innovation* and *activism*, suggesting the public's role in demanding action.

Finally, the dominant interpretants in climate-related headlines focus on immediate threats, the need for urgent action and the potential for innovation. This may indicate that climate news is framed to create a sense of urgency and the need for immediate action, thus contributing to increasing public anxiety. It is worth reiterating that the issue is not that climate change is not serious or urgent, or that people should not be aware of it. The concern is with the tone of the headlines that may be causing anxiety in this population. The power of the Internet to reach the general public cannot be ignored, nor can its power to influence human behavior [8]. Then it is not unlikely that a whole population will be exposed, intentionally or not, to psychologically harmful content. An excess of input with a stressful tone can trigger situational anxiety. Emphasize that this is different from a mental health concern during a disaster, but a concern with mental collapse as a disaster in itself.

For a thorough semiotic analysis it is also necessary to consider the interpretant triplets in order to understand how semiosis is working, Table 5.3 shows the three most common triplets in the dataset. As can be seen, the number of recurring interpretant triplets is too small to provide a general understanding (there are 1407 unique triplets in the dataset). This is actually to be expected given the amplitude of the domain being addressed. Perhaps in a narrower domain, or with a sufficiently large dataset, possibly covering a wider time span, the patterns would emerge.

Nevertheless, it is curious that all three triplets (at least 2% of the sample) resulted in a call to action rather than a pessimistic semiosis. Therefore, one way of extracting insights that may be appropriate to this dataset is clustering. Given the scope of this book, the first thought is to ask the LLM to cluster the data. The difficulty lies in the need to include all elements in a single prompt. LLMs, even those with a sufficiently large context window, are not good at handling tasks with large lists. In the response, some items are usually missing, and others are slightly changed, so the result is not as reliable as it needs to be for this stage of analysis. It is not yet clear whether this is

Table 5.3 Common Patterns of Interpretants for the different datasets

Source	Immediate	Dynamical	Final	Occurrences
Google	Infrastructure	Investment	Sustainability	10 (1.0%)
	Protest	Activism	Disruption	8 (0.8%)
	Protest	Activism	Urgency	6 (0.6%)
Bing	Protest	Activism	Disruption	6 (0.8%)
	Protest	Activism	Urgency	5 (0.7%)
	Disaster	Devastation	Humanitarian Crises	4 (0.5%)
Both	Protest	Activism	Disruption	14 (0.8%)
	Protest	Activism	Urgency	11 (0.6%)
	Infrastructure	Investment	Sustainability	10 (0.6%)

a limitation of the LLMs or whether it is due to the lack of a benchmark of this type that would drive development in this direction. An algorithmic approach is needed.

This situation is an example of how LLM-based and "traditional" NLP methods merge to meet NLA needs, and why one cannot do without the other. Then, in short, the triplets are clustered using Affinity Propagation *cf.* [9], resulting in 205 clusters with fairly well distributed elements (skewness 0.58). For visualization, the Jaccard distance between the $Q1$ triplets (55 clusters with more than 9 triplets, $\approx$ 25% of the sample) is computed, resulting in the cluster-map shown in Fig. 5.4. The option for the Jaccard distance (lexical) instead of embedding distance (semantic) is for reassuring the position that all existing types of NLP approaches may be used for conducting the analysis as far as it makes sense for the task in hand.

For an analytical highlight, a region with high proximity between clusters is highlighted in the figure. This can be called as the "sustainability region" (analysis of other regions are omitted), suggesting that a significant part of the titles in the dataset articulate sustainability-related concerns. In turn, it is possible to see that *sustainability* is often articulated together with *innovation, adaptation, responsibility* and *activism*. All these interpretants imply personal behavior change, suggesting that *sustainability* has the same implication. As behavior change is an intrinsic stressor, it suggests that titles related to *sustainability* may also be a source of climate anxiety.

It is important to emphasize that the analysis and results following this approach are supposed to be understood in the context of the bias present in the model (in the case of this illustration, Llama3 : 8B). This means that the interpretations and insights generated reflect the predispositions inherent in the model, shaped during its training. This intentionality does not necessarily determine whether the results are correct or incorrect, but it does provide a relevant perspective that can influence analysis and decision-making within the context established by the model. The reader is invited to explore Fig. 5.4 further for additional insights it may provide.

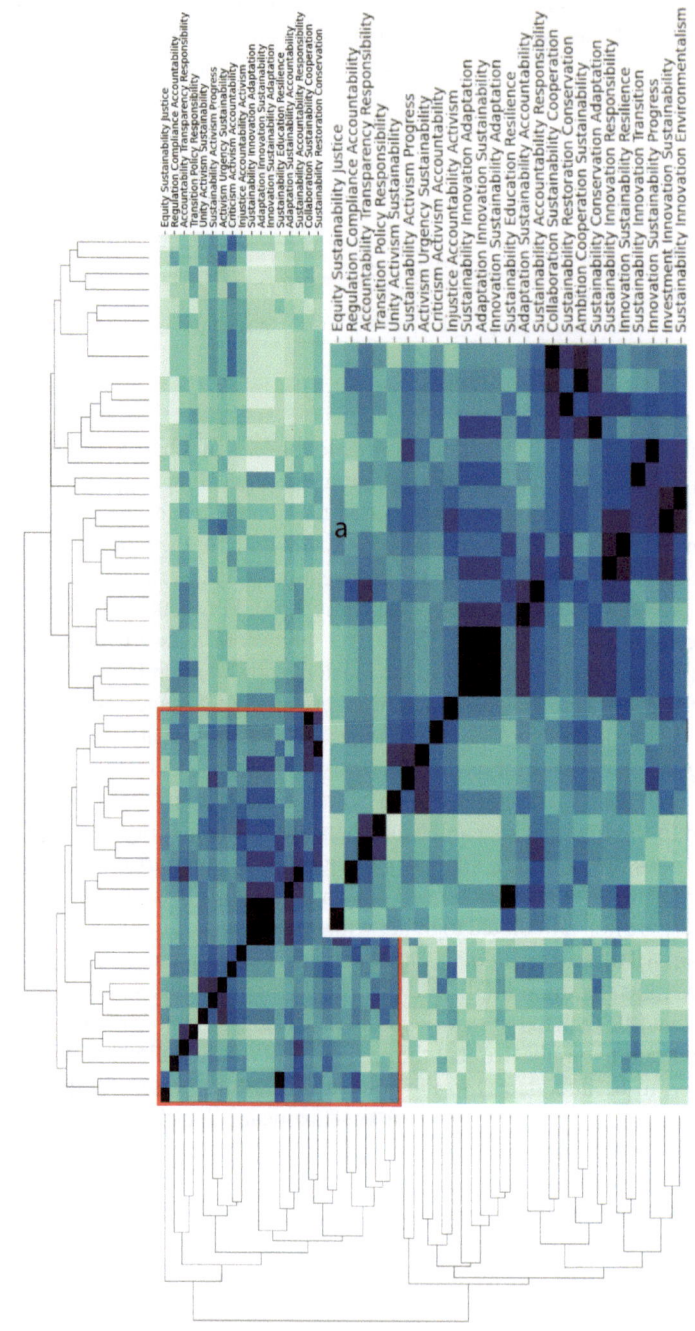

Fig. 5.4 Cluster-map for Jaccard distance of interpretant triplets (x-ticks removed due to space constraints, darker is closer)

5.4 Soft-Data Indicator

Based on the data presented, it is possible to propose a KSI. Indicators play a central role in decision-making by presenting complex information in a straightforward format. This helps decision-makers make a quick and informed response without getting lost in the details, supporting then an effective call to action [10].

The quality of any indicator depends on the soundness and representativeness of the data on which it is based. Indices are often linked to a time frame that allows the creation of time series to track the evolution of the index over time. The resulting score in this example can be viewed as a record within the assumed time series, with each record corresponding to a weekly set of climate data.

The proposal of KPIs often implies a numerical score (e.g. [0..1]), often associated with a textual scale (e.g. low to high). The KSI is no different, except that whereas the KPI is based on hard data (objective), the KSI is based on soft data (subjective). Note that the soft data for the KSI does not necessarily come from an LLM, it can also be provided by human scorers. Consider the table 5.1, any person is able to score the feeling of anxiety caused by the interpretants [11]. By tabulating the data provided, the index is obtained.

The point is that it is possible for an LLM to score words or phrases [5] in the same way that it associates them with types. Prompt 5.6 depicts an example of such an approach. Which is preferred for this case study.

Prompt 5.6 `Phi3:mini-128k`

```
1  >>> Score the anxiety for the list of interpretants: 1)
   ↪  Immediate = Urgency; 2) Dynamic = Activism; and 3) Final
   ↪  = Resilience. Combine the three interpretants into a
   ↪  single anxiety severity score between 0 and 10 (or very
   ↪  low and very high). The immediate interpretant reflects
   ↪  the initial interpretive impact. The dynamic interpretant
   ↪   is the long-term impact. The final interpretant
   ↪  represents resolved states of mind. Output only the score
   ↪  , with a very brief explanation of the values assigned to
   ↪   each interpretant and the calculation procedure.
2
3  Immediate = Urgency: Score - 8 (The immediate fears feel highly
   ↪  urgent)
4  Dynamic - Activism: Score - 5 (Long-term activism is motivating
   ↪  but not as anxiety provoking)
5  Final = Resilience: Score - 6 (A sense of resilience provides
   ↪  some relief from ongoing concerns, though still
   ↪  significant worries exist.)
6
7  Calculation Procedure:
8  Total score calculation uses weighted averages to reflect the
   ↪  varying impacts. Assigning weights based on personal
   ↪  perception and psychological understanding yields an
   ↪  overall anxiety severity score for this particular case
   ↪  of 6 out of ten (7/10 after rounding). The immediate
   ↪  fears dominate, but with some hope due to long-term
   ↪  activism.
```

Table 5.4 Instances for the overall result

Title	Turning the Tide: State of Climate Emergency 2024		
Interpretants	Urgency	Activism	Sustainability
Anxiety Score (single)	High (0.7)	Very High (0.9)	High (0.75)
Anxiety Score (triplet)	Very High (0.95)		
(a) First record			
Title	Why you should prepare for climate disasters with your neighbours		
Interpretants	Preparedness	Community	Resilience
Anxiety Score (single)	High (0.7)	Very Low (0.2)	High (0.75)
Anxiety Score (triplet)	High (0.75)		
(b) Second record			

It is then possible to further process the generated interpretants into a score, resulting in the Anxiety Climate Index (ACI). This process results in a scalable and quantifiable form of the subjective nuances expressed by the interpretants mapped in the context of climate anxiety. It should be emphasized that, as an illustrative example, this is a simplified version of what would be expected from an actual, properly validated, industry-level index.

For this illustration, ACI ranges from very low to very high, and the prompt is designed to assign a score to the generated level after linguistic mapping, where very low means a score close to zero and very high means a score of about one. The model is prompted to give more weight to the immediate interpretant, as it reflects the initial interpretative impact, followed by the dynamic interpretant, which has a significant but lower weight due to its long-term impact, followed by the final interpretant, which has the lowest weight, as it may represent resolved states of mind. This is done for each interpretant, but also for the triplet independently (this discussion focuses on the anxiety score for the triplet). Two examples with the results of the two analysis steps are shown in the table 5.4. ACI's `modelfile` is presented in the prompt 5.7.

Prompt 5.7 ACI `modelfile`

```
1  FROM llama3.1:8b
2
3  PARAMETER temperature 0
4
5  SYSTEM """
6  # Task
7  You are an agent that classify interpretants into levels of
        ↪ anxiety in the context of climate change. The possible
        ↪ levels of anxiety for classification range from very low
        ↪ to very high, where very low equals a score of 0.1, and
        ↪ very high equals a score of 1. Knowing that the immediate
        ↪  interpretant has more weight because it reflects the
        ↪ initial interpretative impact, followed by the dynamic
        ↪ interpretant, which has a significant but lesser weight
        ↪ because of its long-term implications, followed in turn
```

```
   ↪  by the final interpretant that has the least weight as it
   ↪    may represent resolved states of mind. After classifying
   ↪    and scoring each interpretant, merge the three
   ↪  interpretants classifications and scores into a single
   ↪  anxiety level and a single anxiety score, with the
   ↪  results also in the range between very low and very high,
   ↪    and score between 0.1 and 1.
8
9  # Output format
10 After you classified each interpretant anxiety level and scored
   ↪  for anxiety each interpretant, and merged the three
   ↪  interpretants classifications into a single anxiety level
   ↪  , and a single anxiety score, provide the result in the
   ↪  following format, without any additional information:
11 Print: <Immediate Interpretant anxiety level and score>, <
   ↪  Dynamical Interpretant anxiety level and score>, <Final
   ↪  Interpretant anxiety level and score>, <Merged
   ↪  Interpretants anxiety level and score>"""
```

This is a deliberately simpler prompt, and although the model generated scores for all interpretants, it skipped generation for 168 ($\approx$ 10%) triplets. For this case study, the NaN rows were excluded from the final scoring of the index. Finally, the climate of fear generated by the considered search engines in the last week of July 2024 is 73% $\pm$ 2% (confidence interval for the mean of 95%) within the interval [0.06..0.99] (see Fig. 5.5 for a visualization). Therefore, according to the proposed scale, the ACI level at this time is of *moderately high* anxiety. Therefore, as a result that directly follows the previous analyses, the titles collected for the working dataset contribute to the increase of climate anxiety on the Internet.

It should be noted that ACI is only one KSI that would be an indicator on a full dashboard. Another that is also expected is an emotion indicator. This can be achieved by replacing `mindset` in the prompt 5.1 with `emotion` and updating the few-shot. This results in 160 emotions with a prevalence of *alarm* (13%), *concern* (13%) and *hope* (11%). This suggests that despite the anxiety-provoking nature of the topic, there is an undercurrent of optimism or positive anticipation. Note that this is consistent with the idea of "constructive urgency" left by the ACI.

In addition, there are several other tasks that would be of interest to a dashboard as such. Named Entity Recognition (NER) is one of them. In short, it is worthwhile to understand, for example, which entities are associated with higher and lower levels of climate anxiety. After obtaining the entities using SpaCy, the importance of human agency and global cooperation became apparent. Human-related entities such as "activists", "Harris", "Biden" and "Trump" are frequently mentioned. In addition, the global nature of climate issues is underlined by the inclusion of diverse locations such as "South Africa", "Paris", "India" and "California". Taken together, these findings show that the problem requires a coordinated global response that includes government action, environmental protection and sustainable development.

As already discussed, it is expected on NLA that the whole NLP toolbox to be employed to get the most useful result. Note that LLM can also be used to extract NER, resulting in an enriched entity set to support additional insights, see Fig. 5.5.

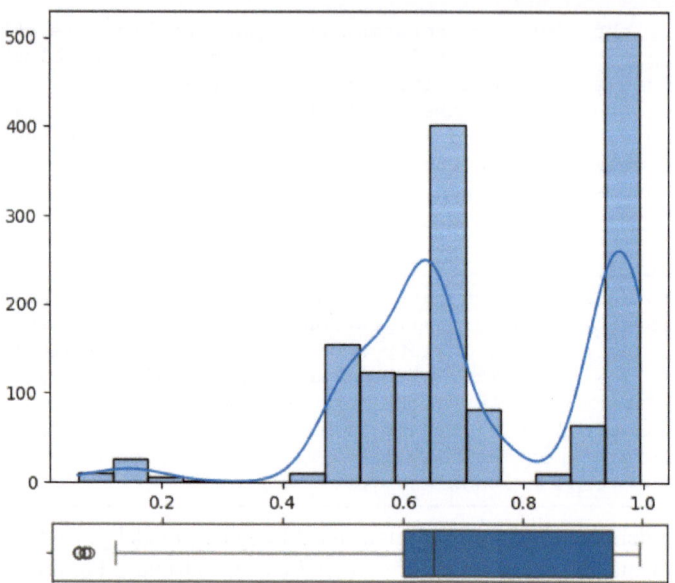

Fig. 5.5 Description of Anxiety Score KSI data

Table 5.5 Named entity recognition

Entity	Count	Entity	Count
ORG	375	GOVERNMENT	80
GPE	284	POLITICS	60
PERSON	166	PERSON	60
DATE	108	PROTEST	54
LOC	74	SUSTAINABILITY	51
MONEY	65	ENERGY	50
CARDINAL	59	FINANCE	44
EVENT	28	ECONOMY	39
ORDINAL	11	WEATHER	38
LAW	7	POLLUTION	32
(a) SpaCy NER		(b) Llama3.1:8b top-10 NER	

It may be the case that both sets are produced for a complementary picture. "Traditional" NER often conforms to a specific and widely used list of types into which the model will attempt to fit the text. LLM-based NER, on the other hand, provides a fine-grained list, but generates a larger number of types. So there is a trade-off to consider. For example, while traditional NER might recognise *Canada* and *Climate Change*, the non-specific model also identifies related terms such as *Climate Tech Startups* and *Climate Change Bill*.

Sidenote on other Generative Tasks

As perhaps the most common task on LLMs is writing support, although this is beyond the scope of this book, a word may be worthwhile. The working claim for the case study is that headlines can contribute to increased climate anxiety by loading a high anxiety weight onto their text. As the population needs to be continuously exposed to content on this topic, it could be journalistically responsible to avoid situational anxiety by reducing psychological load as much as possible, and to prevent this population from becoming habituated to anxiety-laden messages by maintaining a cognitive reserve. In addition to analytics, dashboards and alert systems, LLMs can also help improve communication with a target audience by generating messages with varying levels of urgency to be used according to the severity of the situation. For a glimpse into the use of semiotic prompts to create messages tailored to different psychological profiles, see the prompts 5.8 and 5.9 (note that this is an exploration of possibilities, not an actual, properly crafted and evaluated solution).

Prompt 5.8 Weather Wraning SYSTEM Prompt

```
1  # Task
2  You are an agent that operates as a generative semiotic
       ↪ communication system. Create a set of engaging tweets for
       ↪  spreading the warning based on the description of the
       ↪ weather conditions and the target mindset model. The goal
       ↪  is to inform and keep people safe.
3
4  # Chain-of-thought Analysis Model
5     {omitted, refer to Prompt 5.1}
6
7  ## Mindset Model
8  Find the Mindset based on the Interpretant:
9  - Sentiment: <fit the interpretant> deciding [positive, neutral
       ↪  or negative]
10    - If Interpretant's Sentiment is positive, return the Mindset
          ↪  for the tweet: (high extraversion, high agreeableness
          ↪ , high openness)
11    - If Interpretant's Sentiment is neutral, return the Mindset
          ↪  tor the tweet: (modium agreeableness, medium
          ↪ conscientiousness, medium neuroticism)
12    - If Interpretant's Sentiment is negative, return the Mindset
          ↪  for the tweet: (low agreeableness, high
          ↪ conscientiousness, high neuroticism)
```

Prompt 5.9 Warning Tweets (this is a real message from the Civil Defence in Hawaii.)

```
1  >>> The National Weather Service forecasts heavy rain, flahs
       ↪ flooding, high surf, and the possibility of thunderstorms
       ↪  for Hawaii Island starting Saturday. Residents in flood-
       ↪ prone areas are advised to be on the alert for possible
       ↪ flooding conditions.
```

```
2
3   Here are three tweets, each targeting a different mindset:
4   ### Example 1: Negative Mindset
5   > Tweet: "RED ALERT! Heavy rain, flash flooding, high surf, and
          ↪ thunderstorms are coming to Hawaii Island! If you're in a
          ↪  flood-prone area, take immediate action to protect
          ↪ yourself and your loved ones. Stay safe! #FloodWarning"
6   ### Example 2: Neutral Mindset
7   > Tweet: "Heads up, Hawaii Island! The National Weather Service
          ↪ is forecasting heavy rain, flash flooding, high surf, and
          ↪  thunderstorms starting Saturday. Stay informed, stay
          ↪ prepared, and stay safe! #WeatherAlert"
8   ### Example 3: Positive Mindset
9   > Tweet: "Aloha, Hawaii Island! Let's come together to prepare
          ↪ for the severe weather forecasted for this weekend. Check
          ↪  in on your neighbors, stock up on supplies, and stay
          ↪ informed. We got this! #CommunityFirst"
```

In sum, the exploration of the ACI illustrates how to extract and transform qualitative text-data into KSIs. Through the case study, it has been shown that LLMs, when integrated into downstream NLA processes, may uncover "hidden" patterns and provide a nuanced view of data. The next chapter is to discuss the broader implications of using LLMs and NLA, not only for individual case studies, but for systemic applications across different domains for concluding remarks.

References

1. R. Van Solingen, V. Basili, G. Caldiera, and H. D. Rombach. Goal question metric (gqm) approach. *Encyclopedia of software engineering*, 2002.
2. K. Andersen, M. Djerf-Pierre, and A. Shehata. The scary world syndrome: News orientations, negativity bias, and the cultivation of anxiety. *Mass Communication and Society*, 27(3):502–524, 2024.
3. R. Maini, L. Clarke, K. Blanchard, and V. Murray. The sendai framework for disaster risk reduction and its indicators-where does health fit in? *International Journal of Disaster Risk Science*, 8:150–155, 2017.
4. R. Plutchik. The nature of emotions: Human emotions have deep evolutionary roots, a fact that may explain their complexity and provide tools for clinical practice. *American scientist*, 89(4):344–350, 2001.
5. F. Marcondes, A. Gala, M. Rodrigues, J. J. Almeida, and P. Novais. Lexicon annotation with llm: a proof of concept with chatgpt. Paper submitted to a Conference., 2024.
6. H. W.J. Rittel and M. M. Webber. Dilemmas in a general theory of planning. *Policy sciences*, 4(2):155–169, 1973.
7. A. Gala. *Confrontações entre máquinas físicas, máquinas semióticas e máquinas ontológicas*. PhD thesis, Pontifícia Universidade Católica de São Paulo, 2016.
8. B. Lindström, M. Bellander, David T. Schultner, A. Chang, P. N. Tobler, and D. M. Amodio. A computational reward learning account of social media engagement. *Nature communications*, 12(1):1–10, July 2021.
9. B. J. Frey and D. Dueck. Clustering by passing messages between data points. *Science*, 315(5814):972–976, 2007.

10. T. H. Davenport. Analytics 3.0. *Harvard business review*, 91(12):64–72, 2013.
11. S. Mohammad. Best practices in the creation and use of emotion lexicons. In Andreas Vlachos and Isabelle Augenstein, editors, *Findings of the Association for Computational Linguistics: EACL 2023*, pages 1825–1836, Dubrovnik, Croatia, 2023. Association for Computational Linguistics.

Chapter 6
Conclusion

TRIPLE BAM!!!

Josh Starmer

Abstract This chapter revisits key themes and underpins the contributions of this book in a broader sense. The discussion includes the evolution of LLMs, from early developments to the widespread adoption of ChatGPT and other models, highlighting their impact on research and applications. The chapter summarizes the transformative potential of LLMs in NLA, emphasizing their role in supporting qualitative insights by extracting soft data, which is crucial for nuanced decision making. It also provides a rationale for emerging techniques such as fine-tuning, enrichment strategies, and the importance of addressing environmental, legal and ethical considerations. The chapter concludes by discussing future directions and implications of LLMs in both organizational and societal contexts, encouraging further exploration and responsible innovation in this rapidly evolving field.

Revisiting the authors journey, the contribution of this book is a result of a convergence of three different fields of knowledge that the authors have been wrestling with for some time: (1) **cognitive semiotics**: the study of models of feelings, emotions, mindsets, and culture in the context of semiotics; (2) **NLP**: research mainly within lexical and syntactic paradigms; and (3) **big data**: analytics, machine learning and institutional decision-making. Before the introduction of ChatGPT a couple of years ago, the plasticity of cognition and linguistics appeared nearly irreconcilable with the rigidity of hard data from analytics and computational methods based on lexical and syntactical solutions.

Although significant progress had been made with GPT-2 and GPT-3, it was ChatGPT, a variant of GPT-3 fine-tuned with reinforcement learning with human feedback (RLHF), that gained widespread popularity and accessibility. This model set a new standard and brought a new pace and breath to research.

ChatGPT started a battle between big techs like Google, who released Bard (currently Gemini), and Meta, who released Llama. At that early stage, LLMs without

F. S. Marcondes et al., *Natural Language Analytics with Generative Large-Language Models*, SpringerBriefs in Computer Science,
https://doi.org/10.1007/978-3-031-76631-2_6

content moderation, some of them running on the dark web, such as the `WormGPT` and `FredomGPT`, were also delivered. Unlike `ChatGPT` and `Bard`, which were proprietary tools, `Llama` was perhaps the first open-source model, sparking a flurry of research based on it. Tools such as `llama.cpp` and *Ollama*, as well as concurrent and derived models such as `TinyLlama`, `Alpaca` and `Llava`, are named after it.

Other players appeared, resulting in new open source and proprietary foundation models such as `Qwen`, `Gemma` and `Claude`. Finally, as this book is being written, Meta has released `Llama3:450B` that matches the performance of `ChatGPT-4o` on several benchmarks. This has stimulated all sorts of research into the subject by both researchers and technicians. This led to a renewed set of relevant improvements, such as better forms of fine-tuning, ways of reducing hardware requirements, strategies for improving factual knowledge, *etc*. Some researchers, as well as nearby professionals and companies, opened up spaces for implementing and testing these solutions, believing in the potential for scientific innovation on the one hand and business process improvement on the other.

Considering the fields of knowledge pursued by the authors, all these events became related to the field of *NLP*. Being aware of these developments at the same time with extensive experiments, the bridge between LLMs and *cognitive semiotics* was established (both theoretically and empirically) and the notion of *soft data* was formalized. With analytics related to sentiment analysis and emotion extraction already widespread [1, 2], the convergence with *big data* towards *natural language analytics* was the next natural step. Solutions for processing and analyzing large data sets, such as the Python interface with Ollama, provided the backdrop for scaling qualitative linguistic data, leading to concepts such as Key Soft Indicators. This is where the book makes its main adding to the field.

6.1 Synthesis and Discussion

As a synthesis, this book proposed and explored the transformative potential of LLMs in the context of NLA. It highlights the practical utility of LLMs in extracting qualitative data, referred to as "soft data", from semantic tokens, emphasising their importance beyond mere accuracy. LLMs are metaphorically compared to language calculators, highlighting their ability to semiotically drive and perform subjective analysis similar to a human's evaluation, which in turn is essential for nuanced decision-making in organizational contexts. It suggests an extra dimension to decision-making, adding possibilites to a process that is still largely driven by hard data.

The exploration extended to the underlying principles and practical applications of NLA, with discussions on how machine learning algorithms, in particular the Transformer architecture, use large text corpora to generate embeddings that capture semantic relationships. Embeddings that enable the understanding of language nuances by taking into account their dynamics in vector space, enabling the extraction of valuable insights from text data into KSIs. KSIs complement traditional

KPIs, highlighting and reinforcing the critical role of qualitative data in organisational decision-making, supported by NLP tasks such as Named Entity Recognition, Relation Extraction and Semiotic Analysis.

Other relevant topics also presented are the methodologies of prompt engineering, a central aspect in aligning LLM outputs with user-specific needs and mitigating or exploring inherent biases. By categorizing prompting techniques and incorporating semiotic principles, it illustrated how structured prompts can enhance the precision and contextual relevance of LLM responses. Practical tools like Ollama and a case study on the application of LLMs in a climate anxiety scoring was presented to show some practical use cases. Additional topics that yet require further discussion are *fine-tuning* and *enrichment*. Both are strategies for improving the accuracy of the LLM response, the first from inside and the second from outside. In short, fine-tuning adjusts the weights of the model to match the expected results of a given dataset. Enrichment, in turn, provides additional context and information to help the model generate a better response. As a result, the former biases the generation towards a particular style [3] and the latter increases the model's recall on a particular topic [4]. From an NLA point of view, these proposals raise relevant issues that are worth discussing in the following sections.

6.1.1 Parameter Efficient Fine-Tuning

Full fine-tuning of an LLM is rarely possible on corporate (and often university) computers due to the computational resources required. Therefore, parameter efficient fine-tuning (PEFT) is being explored, the idea being to update some of the weights while keeping the rest "frozen". Three common approaches are to update the weights: of the input layer (*e.g. prompt tuning*), of the output classifier (*e.g. feature extraction tuning*), or of some parameters in the main model (*e.g. low-rank adaptation*, LoRA). It is worth mentioning that even in situations where the training process would take some hours or days before being ready, the process may be still feasible if enough time is available (note that good performance may not always be achieved at the first attempt). Thus, the PEFT of an LLM on average hardware is a fairly achievable task.

Perhaps the most important criticism of fine-tuning LLMs is that the benefits diminish as the models become more accurate. For example, LoRA produced a remarkable result in improving GPT2 [5]. For the sake of argument, suppose that fine-tuning improved the accuracy of a low-fidelity model from 40% to 60%, what would be a significant gain. Now suppose a base model whose accuracy on the same task is about 80%, a fine-tuning that improves the model to about 5% would also be considered a significant achievement. The point here is that as the performance of a model increases, the potential benefit of fine-tuning decreases. Given that the performance of LLMs on various tasks is continually increasing, the need for fine-tuning may become a disputed issue. Therefore, before embarking on a fine-tuning process, it is advisable to evaluate the current performance of the model on the intended task.

This leads to another problem: it is not always easy to evaluate the performance of an LLM. One difficulty is finding an appropriate metric, and another is obtaining a useful dataset. The same difficulty exists when it comes to fine-tuning, which can then be discussed together. In addition, both evaluation and fine-tuning are usually based on supervised learning, which means that the dataset must contain an input instance paired with the expected prediction. This format is certainly suitable for *hard data*, but not for *soft data*, where some variation in the prediction is expected. Note that the expected variation is not about precision (which could be addressed with Levenshtein or BertScore distances, for example), but about nuance.

Consider the word *coronavirus*, from a scientific point of view it is a neutral word, but from a personal point of view it is likely to be seen as a negative word. These are two nuances that the word can reveal. A common way of annotating subjective data is to use a Likert scale like questionnaire on a crowdsourcing platform such as Mechanical Turk (https://www.mturk.com/). This is, for example, the option provided by both VADER [6] and EMOLEX [7]. Often the annotation, if numerical, is the average of all the scores and, if categorical, the most common category scored by the reviewers. It is a statistical approximation, sometimes under-sampled, of a subjective evaluation. For example, suppose that the mean sentiment score of *coronavirus* is -0.25 (on the scale $[-1, 1]$), meaning that the word has a neutral to negative sentiment. This may not accurately reflect reality; the subjective feeling about *coronavirus* is either totally negative or totally neutral, according to the context. The loss of information is greater if only the most common nuance is used.

Semiotically, it is expected the model to predict a hypothetical interpretant. As discussed, it is not possible to ultimately assert the intention of the writer only considering a given text. The meaning produced by a sign is to be considered within a meaning horizon. Consider the sentence *Sarah buys a new electric car*, the mental trait is considered *agreeableness* to the extent that the cultural horizon considered is *environmentally conscious*, considering a cultural horizon such as *adoption of new technology* the trait would be *openness*. Both are correct in the sense that they are possible nuances of a same phenomenon, but from different cultural perspectives.

In this sense, it is an open question how to create soft data datasets that can be used either for evaluation or for fine-tuning. Finally, this explains why this book does not use fine-tuning.[1], preferring the *hard prompt*[2] associated with a specific model setup.

One insight that may follow in this context, derived from hard prompt fitting, is to use the LLM to bootstrap PEFT. In short, the idea is to use an LLM to provide classifications for a dataset, which would be revised and cleaned by a specialist and

---

[1] Information on how to perform PEFT can be found at HuggingFace: https://huggingface.co/PEFT.

[2] Hard prompt is a term probably derived from hardcoding, which is the process of writing, usually configuration parameters, directly into the source code. Thus, "hard-prompting" is the process of writing the prompt "by hand", as opposed to *soft prompt*, a machine learning fitted prompt, which is the result of prompt tuning.

then used to fine tune one's model. The intention is to stabilize the taxonomy of the output according to the tendencies already present in the model. However, this is an over-complicated approach, as it would require fine-tuning the model for several parameters at the same time. A simpler approach would be to create a classifier for each parameter and use it to post-process the dataset.

6.1.2 Contextualized Information Enrichment

Perhaps the most common approach in contextualized information enrichment is Retrieval Augmented Generation (RAG) [8]. In a nutshell, a database of embeddings extracted from a given source is compared with the embedding of the query, with the aim of finding the most semantically similar block of sentences. Some of the most similar blocks are fed into the LLM, along with the original prompt and additional instructions. The LLM produces an answer based on the data provided. Another related approach is Mixture-of-Agents (MoA) [9]. Again, in short, it is an ensemble of different LLMs (often each with its own specialization) that produce a response to the same prompt. The weighted responses are fed into the prompt of the core LLM, which produces the final response based on the responses.

Emphasize that this strategy does not require a dataset, but does require external sources from which data can be queried. As far as could be seen, at least simple RAG is not suitable for NLA. Consider, for example, an emotion classification engine based on Plutchick's theory. There is no reason to believe that retrieving paragraphs from Plutchick's bibliography that are semantically similar to the user query would improve the classification of that input.

MoA, on the other hand, would fall in the same case of multiple people annotating a dataset. As already discussed, if the goal is to obtain an unambiguous classification, it does not make sense to search for means or modes in soft data. As also discussed, a suitable approach is to have a specialist create the prompt of a single model and place it in the desired meaning horizon for getting useful insights. If the aim is to obtain insights from different sources that can be considered together, despite not being MoA, is achievable. This may be particularly relevant when considering foundational models trained in different languages [10].

6.2 Related Considerations

After a winter period, the field of artificial intelligence (AI) begin to expand at a rapid pace, bringing with it a set of considerations that are relevant to be mentioned to ensure its responsible development and deployment. This section briefly presents some considerations, focusing on the ecological footprint, legal constraints, and mentioning economic and social impacts of AI technologies, particularly of ChatGPT (still the most famous, then approached LLM).

6.2.1 Ecological Footprint

As most proprietary models do not disclose details of their development or use, it is only possible to estimate their ecological footprint. For this reason, there are only a few papers on the subject, with the carbon footprint being discussed in [11] and the water footprint in [12]. In summary, training AI models such as GPT-3 requires huge amounts of water, mainly for cooling energy-intensive data centers. Even the inference process, in which the trained model is used, requires a significant amount of water (approximately 500 ml of water per 5 to 50 prompts). Training AI models also contributes to significant carbon emissions due to the high energy consumption of data centers. The level of carbon emissions varies depending on the fuel mix used to generate electricity at the data center location. It should be noted that data centers located in regions with high carbon electricity sources and in areas with water stress contribute to a higher carbon footprint and increased water consumption respectively. Therefore, the geographic location of data centers plays a critical role in determining the overall environmental impact of AI operations. The environmental impact of AI models such as ChatGPT goes beyond carbon emissions and water consumption. The production and disposal of hardware, the sourcing of rare materials, and the long-term sustainability of data center infrastructures also contribute to the environmental footprint.

While specific data on the environmental footprint of ChatGPT is limited, the lessons learned from studies of GPT-3 and similar models underscore the importance of addressing the environmental impacts associated with AI. For example, a side effect of running quantized LLMs locally, especially when considering an enterprise-level deployment environment based on CPUs or small GPUs, helps to reduce the ecological footprint, at least during the prediction phase. Currently, massive models are being trained one at a time, but it is expected that this will eventually stop as the performance of the models become similar. At this stage of maturity, the training footprint is expected to decrease, but the prediction footprint is expected to increase continuously, considering that these tools may eventually become ubiquitous in society. In this situation, reducing the footprint by running the LLMs on local, energy-efficient computers can be a cornerstone to this technology long-term viability.

6.2.2 Legal Constraints

As any human activity, AI is also regulated, at least in Europe, through the AI Act and the General Data Protection Regulation (GDPR). Needless to say that any initiative must comply with local laws. The key concern of these initiatives is to protect individual rights. In addition, these laws are somehow linked. For an instance, the AI Act forbids the use of AI to perform social scoring which, in turn, would be only possible with sufficient personal data of an individual. Therefore, it is permissible to make judgements about film descriptions as shown in Fig. 2.1. On the other hand, it is not allowed to infer the personality of an individual without their knowledge in order to mediate interactions with them. Therefore, being in accordance with the GDPR, it is allowed to extract such information for statistical purposes, *i.e.* for analytics. As a basis for discussion, see an extract from the AI Act:

AI Act, Chapter 5

1. The following AI practices shall be prohibited:..

   (c) the placing on the market, the putting into service or the use of AI systems for the evaluation or classification of natural persons or groups of persons over a certain period of time based on their social behaviour or known, inferred or predicted personal or personality characteristics, with the social score leading to either or both of the following:

      (i) detrimental or unfavourable treatment of certain natural persons or groups of persons in social contexts that are unrelated to the contexts in which the data was originally generated or collected;
      (ii) detrimental or unfavourable treatment of certain natural persons or groups of persons that is unjustified or disproportionate to their social behaviour or its gravity;

For the full description of these legislations, please follo+w:

GDPR   https://gdpr.eu/
AI Act   https://artificialintelligenceact.eu/

6.2.3 Economic Effects and Social Impact

Needless to say, the use of LLMs is expected to have significant economic and social impacts. The deployment of LLMs is expected to reshape various industries, influence job markets, and alter social dynamics. These transformations bring both opportunities and challenges that warrant further discussion and research. As these issues are extensively discussed in the media and other researches, and also as the relationship with this book is ancillary, the discussion of them will be omitted. The purpose of this section is to highlight the authors' awareness and concern about these issues and to stimulate further discussion on them.

6.3   Summary and Future Trends

This book has provided an in-depth examination of the potential of LLMs to extract actionable insights from textual data, exploring the methods, tools, and implications of using these advanced models for natural language analysis (NLA). Throughout the chapters, we have seen how generative LLMs, particularly through the use of tools such as Ollama, can be used to extract meaningful data from unstructured text. This involves not only traditional NLP techniques, but also advanced prompt engineering and semiotic analysis, allowing for a deeper understanding of the underlying sentiments, emotions, and mindsets within the data.

The book began with a basic overview of LLMs, setting the stage for their application in NLA. It emphasized the importance of understanding both the syntactic and semantic layers of language, which are critical to creating models that are not only accurate but also aligned with human intent. The practical chapters demonstrated the use of these models in real-world scenarios, such as analyzing climate change narratives to develop an Anxiety Climate Index (ACI). This case study illustrated the power of LLMs to transform soft data into actionable insights, providing decision makers with valuable tools to address complex societal issues.

The journey through this book has revealed both the opportunities and challenges associated with the use of LLMs in different domains. On the one hand, the ability to process and analyse vast amounts of textual data is opening new avenues for innovation, particularly in fields that require a nuanced understanding of human communication, such as marketing, public health, and environmental science. The case studies presented underscore the potential for LLMs to contribute to these fields by providing deeper insights into human behavior and societal trends.

On the other hand, the challenges are equally significant. The environmental footprint of training and deploying these models is a concern that cannot be overlooked. As the demand for AI systems grows, so does the need for energy-efficient solutions that mitigate the environmental impact of these technologies. In addition, legal and ethical considerations, particularly related to data privacy and the use of AI in decision-making, require careful attention. The ongoing development of regulations, such as the AI Act and GDPR in Europe, highlights the importance of ensuring that AI technologies are developed and used responsibly.

Looking ahead, as mentioned above, several trends are likely to shape the future of AI and its applications: ecological footprint, legal constraints, and economic effects and social impact. However, integration with other technologies, continued advances in NLP, a focus on interpretability and trustworthiness, and collaborative and open source AI development will also be important.

The time at which this book is being published is a critical one, as it is at the turning point between the *peak of inflated expectations* and the *trough of disillusionment* in the hype cycle [13]. In short, while rapid progress is being made on a number of fronts, major journals such as The Economist and Forbes are declaring that LLMs are not translating into higher earnings. It remains to be seen whether LLMs will find their way onto the *Plateu of Productivity*. Nevertheless, there is a gap to be bridged between these effects and their profitable use by companies. The authors like to think that this book is a contribution in that direction.

The possibility that LLMs are a technology that is here to stay may be a case of imagining not its near-future consequences, but a bit further ahead. Just as automobiles have shaped the way people get around, the Internet has changed the way knowledge is disseminated and smartphones redefined how communication takes place, so LLMs will shape the way people write, ask questions and perhaps even think. The ongoing evolution of the human brain, driven by these technological innovations, may open new paths for cognitive enhancement and challenge our understanding of human consciousness. This is another exciting (at least for the optimists) line of research that this technology has opened up.

Não tenhamos pressa, mas não percamos tempo.
(Let's not rush, but let's not waste time)
José Saramago (Portuguese Nobel prize)

References

1. W. Medhat, A. Hassan, and H. Korashy. Sentiment analysis algorithms and applications: A survey. *Ain Shams engineering journal*, 5(4):1093–1113, 2014.
2. M. Wankhade, A. C. S. Rao, and C. Kulkarni. A survey on sentiment analysis methods, applications, and challenges. *Artificial Intelligence Review*, 55(7):5731–5780, 2022.
3. A. Gudibande, Wallace, C. Snell, X. Geng, H. Liu, P. Abbeel, S. Levine, and D. Song. The false promise of imitating proprietary llms. *arXiv preprint* arXiv:2305.15717, 2023.
4. Y. Hoshi, D. Miyashita, Y. Ng, K. Tatsuno, Y. Morioka, O. Torii, and J. Deguchi. Ralle: A framework for developing and evaluating retrieval-augmented large language models. *arXiv preprint* arXiv:2308.10633, page 52–69, 2023.
5. E. J. Hu, Y. Shen, P. Wallis, Z. Allen-Zhu, Y. Li, S. Wang, L. Wang, and W. Chen. Lora: Low-rank adaptation of large language models, 2021.
6. C. Hutto and E. Gilbert. Vader: A parsimonious rule-based model for sentiment analysis of social media text. In *Proceedings of the international AAAI conference on web and social media*, volume 8, pages 216–225, 2014.
7. S. Mohammad and P. Turney. Emotions evoked by common words and phrases: Using mechanical turk to create an emotion lexicon. In *Proceedings of the NAACL HLT 2010 workshop on computational approaches to analysis and generation of emotion in text*, pages 26–34, 2010.
8. P. Lewis, E. Perez, A. Piktus, F. Petroni, V. Karpukhin, N. Goyal, H. Küttler, M. Lewis, W. Yih, T. Rocktäschel, et al. Retrieval-augmented generation for knowledge-intensive nlp tasks. *Advances in Neural Information Processing Systems*, 33:9459–9474, 2020.
9. J. Wang, J. Wang, B. Athiwaratkun, C. Zhang, and J. Zou. Mixture-of-agents enhances large language model capabilities. *arXiv preprint* arXiv:2406.04692, 2024.
10. F. S. Marcondes, P. Oliveira, P. Freitas, J. J. Almeida, and P. Novais. he moral dilemma of computing moral dilemmas. 5th International Workshop on Autonomous Agents for Social Good (AASG 2024), in conjunction with the 23rd International Conference on Autonomous Agents and Multiagent Systems (AAMAS 2024), 2024. https://panosd.eu/aasg2024/papers/AASG2024_paper_3.pdf.
11. P. Li, J. Yang, M. A. Islam, and S. Ren. Making ai less" thirsty": Uncovering and addressing the secret water footprint of ai models. *arXiv preprint* arXiv:2304.03271, 2023.

12. I. d'Aramon, B. Ruf, and M. Detyniecki. Assessing carbon footprint estimations of chatgpt. In *Renewable Energy Resources and Conservation*, pages 127–133. Springer, 2024.
13. A. Linden and J. Fenn. Understanding gartner's hype cycles. *Strategic Analysis Report No R-20-1971. Gartner, Inc*, page 88, 2003.